RAND NATIONAL SECURITY RESEARCH DIVISION

AF768437

Hypersonic Missile Nonproliferation

Hindering the Spread of a New Class of Weapons

Richard H. Speier, George Nacouzi, Carrie A. Lee, Richard M. Moore

Sponsored by the Carnegie Corporation

For more information on this publication, visit www.rand.org/t/RR2137

Library of Congress Cataloging-in-Publication Data is available for this publication.

ISBN: 978-0-8330-9916-7

Published by the RAND Corporation, Santa Monica, Calif.

© Copyright 2017 RAND Corporation

RAND® is a registered trademark.

Limited Print and Electronic Distribution Rights

This document and trademark(s) contained herein are protected by law. This representation of RAND intellectual property is provided for noncommercial use only. Unauthorized posting of this publication online is prohibited. Permission is given to duplicate this document for personal use only, as long as it is unaltered and complete. Permission is required from RAND to reproduce, or reuse in another form, any of its research documents for commercial use. For information on reprint and linking permissions, please visit www.rand.org/pubs/permissions.

The RAND Corporation is a research organization that develops solutions to public policy challenges to help make communities throughout the world safer and more secure, healthier and more prosperous. RAND is nonprofit, nonpartisan, and committed to the public interest.

RAND's publications do not necessarily reflect the opinions of its research clients and sponsors.

Support RAND

Make a tax-deductible charitable contribution at
www.rand.org/giving/contribute

www.rand.org

Preface

Hypersonic missiles—specifically, hypersonic glide vehicles and hypersonic cruise missiles—are a new class of threat able to penetrate most missile defenses and to further compress the timelines for a response by a nation under attack. Such missiles are being developed by the United States, Russia, and China. Their proliferation beyond these three nations could result in lesser powers setting their strategic forces on hair-trigger states of readiness and more credibly being able to threaten attacks on major powers.

There is probably less than a decade available to substantially hinder the proliferation process. To this end, this report makes specific recommendations for actions by the United States, Russia, and China, as well as by the broader international community.

This report was prepared in 2015–2017 under the sponsorship of the Carnegie Corporation of New York for its project "Disruptive Technologies and the Future of Deterrence." It should be of interest to individuals and organizations concerned with defense technologies, arms control, or nonproliferation.

This research was conducted within the International Security and Defense Policy Center (ISDP) of the RAND National Defense Research Institute. For more information on ISDP, see www.rand.org/nsrd/ndri/centers/isdp or contact the director (contact information is provided on the web page).

Contents

Preface iii
Figures ix
Summary xi
Acknowledgments xv
Abbreviations xvii

CHAPTER ONE
Introduction: What This Report Addresses 1

CHAPTER TWO
Strategic Consequences of Hypersonic Missile Proliferation 7
Principal Characteristics of HGVs 8
Principal Characteristics of HCMs 11
Long-Term Planning Perspectives for HGV and HCM Technologies 15
Strategic Implications of Hypersonic Weapons 16
The Broader Picture of Increased Risk 18

CHAPTER THREE
Ongoing Hypersonic Technology Proliferation 21
Committed Governments 22
R&D in Less-Committed Countries 28
International Cooperation 29
Claimed Reasons for Pursuing Hypersonic Technology 31
Challenges Posed for Controlling Proliferation 32
Summary 34

CHAPTER FOUR
Hindering Hypersonic Missile Proliferation 35
Unilateral Measures 35
Multilateral Measures 37
Potential Export Controls 39
Is the Missile Technology Control Regime Adaptable to Hypersonic Technology? 42
Recommended Items to Control 44

CHAPTER FIVE
Conclusions 47

APPENDIX A
The Hypersonic Flight Regime 49
Introduction 49

APPENDIX B
Survey of Foreign Hypersonic Activity 53
European Union 54
Australia 56
Belgium 60
Brazil 62
Canada 64
France 64
Germany 68
India 70
Iran 76
Israel 77
Italy 79
Japan 81
The Netherlands 84
Norway 86
Pakistan 87
Singapore 88
South Korea 89
Spain 90
Sweden 91

Taiwan 92
United Kingdom 93

APPENDIX C
Technical and Economic Barriers to Hypersonic Systems Development 99
Technical Barriers 99
Economic Challenges 106
Summary of Challenges 107

APPENDIX D
Suggested Export Control List for Hypersonic Technologies 109
Standard Additions to Export Controls 109
Specific Suggestions for Export Controls 110

References 117

Figures

1.1. Generic Concept of an HGV ... 2
1.2. Generic Concept of an HCM ... 3
1.3. Ballistic Reentry Vehicle (RV) Versus HGV Trajectories ... 4
2.1. Typical HGV and MaRV Trajectories ... 9
2.2. HGV Versus RV Terrestrial-Based Detection ... 11
2.3. Destructive Power of a High-Speed Mass as a Function of Speed ... 13
3.1. French LEA ... 23
3.2. Indian-Russian BrahMos II ... 24
3.3. Australian–U.S. HIFiRE Scramjet ... 26
3.4. Japanese HyTEx ... 27
3.5. European LAPCAT II ... 28
3.6. Chinese Mach 4 Missile Exported to Pakistan ... 31
4.1. Illustrative Ranges from Japan ... 40
4.2. Illustrative Ranges from India ... 41
4.3. Illustrative Ranges from Poland ... 41
A.1. X-15 Hypersonic Test Vehicle ... 50

Summary

This report examines the implications of the proliferation of hypersonic missiles and possible measures to hinder it.

Hypersonic missiles can be maneuverable and travel at approximately 5,000 to 25,000 kilometers per hour, or one to five miles per second. In more familiar terms, these missiles fly six to more than 25 times as fast as modern airliners. They fly at unusual altitudes—between a few tens of kilometers and 100 kilometers. These characteristics of high speed, maneuverability, and unusual altitudes make them both challenging to the best missile defenses now envisioned and, until the last minutes of flight, unpredictable as to their targets.

Hypersonic missiles create new challenges to global security. If hypersonic missiles spread into the international market, the existing threats posed by ballistic and cruise missiles would be compounded. As one example, hypersonic missiles, if used against nations with limited strategic forces, might disarm target forces before they can react. This prospect can lead the target nations to set up their strategic forces for "launch on warning"[1] —creating many forms of crisis instability. And because of the difficulties of defending against hypersonic missiles, relatively small hypersonic forces can pose threats against major powers' forward-projected forces, or even deterrent threats against the homelands of major powers.

Two primary types of hypersonic missiles are emerging. Hypersonic glide vehicles (HGVs) are launched by rockets into near space,

[1] *Launch on warning* is defined as a strategy in which a retaliatory attack is launched before incoming missiles have reached their targets.

where they are released and fly to their targets by gliding along the upper atmosphere. They travel at the upper levels of hypersonic speeds and altitudes. Hypersonic cruise missiles (HCMs) are powered all the way to their targets by rockets or advanced jet engines, such as scramjets (supersonic combustion ramjets). They are faster versions of existing cruise missiles. Both missile types may be ready for military use in a decade or less. Because they are maneuverable, both missile types are far more difficult to defend against than legacy ballistic missiles. Moreover, their flight altitude and maneuverability result in less warning as compared with higher-flying ballistic missiles.

Current Development Efforts

Hypersonic missiles are currently being developed mainly by the United States, Russia, and China. Other countries besides these three are also developing hypersonic technology to some degree. France and India are the most committed, and both draw to some extent on cooperation with Russia. In terms of level of effort, the next programs are those of Australia, Japan, and European entities.

Hypersonic technology has a dual-use character; it can be used for nonmilitary purposes including space launch, spacecraft retrieval, and civilian transport of passengers and cargo. However, once a nation acquires hypersonic technology, its intentions can change. The technology can be imported or exported, short-circuiting the slow route of indigenous development. The current situation, with hypersonic research openly disseminated and widely spread among governments, industries, and universities, presents challenges for nonproliferation.

On the other hand, there are formidable technical barriers to mastering such hypersonic technologies: thermal management and materials; air vehicle and flight control; propulsion for HCMs; and testing, modeling, and simulation. In addition, there are serious economic uncertainties about the market for some commercial applications, including hypersonic airliners. All these raise the possibility that, with restraint in international cooperation, the diffusion of hypersonic missiles can be limited.

A Game-Changing Capability

There are strategic considerations in favor of limiting hypersonic missile proliferation. Hypersonic missiles do not necessarily increase the vulnerability of nations that do not have missile defenses; they are already vulnerable to current types of missiles. However, an increasing number of nations are acquiring missile defenses that could be penetrated by hypersonic missiles. A hypersonic attack could occur with very little warning time; this factor and the unpredictability of the targets of a hypersonic attack compress the timeline for response by the party being attacked. Hypersonic missiles also increase the expectation of a disarming attack. These threats encourage the threatened nations to take such actions as devolution of command and control of strategic forces, wider dispersion of such forces, a launch-on-warning posture, or a policy of preemption during a crisis. In short, hypersonic threats encourage hair-trigger tactics that would increase crisis instability. The threat is greatest for nations with limited resources but investments in missile defenses. However, major powers are also threatened by the proliferation of hypersonic missiles and the crises they can exacerbate. The more that hypersonic missiles proliferate into the hands of additional nations, the more paths develop for crises.

Nonproliferation Options

There are, however, measures that can hinder such proliferation beyond the United States, Russia, and China. Unilateral measures, such as classification, unilateral export controls, and attempts to develop defenses, have limited value if other governments decide to export the missiles or their technology. Such traditional international measures as bans on hypersonic missiles can be counterproductive to negotiate and are not necessarily of interest at the current stage of hypersonic weapon development.

The most promising approach appears to be multilateral export controls. If the United States, Russia, and China embargo complete hypersonic missiles and their major subsystems, the proliferation of this difficult technology would be sharply hindered. As with other forms of nonproliferation, this action could be amplified by other like-minded nations—or nations that simply prefer not to have hypersonic missiles in their neighborhoods. Our research suggests that France could play a key role in organizing the international community for such controls.

This research examines specific hypersonic technologies that could be subject to export controls. The model for such controls is the 35-nation Missile Technology Control Regime (MTCR), which already incorporates some controls on hypersonic-related technologies. However, the MTCR aims to inhibit only the proliferation of missiles capable of delivering nuclear, chemical, or biological payloads, and hypersonic missiles need not deliver a mass destruction warhead in order to be effective. So export controls on hypersonic missiles may require some policies outside of the MTCR or hybrid approaches within and outside of the regime.

Recommendations

Within the structure of MTCR-type controls, this report outlines a two-tiered approach to containing the spread of hypersonic systems and components. First, we recommend a policy of export denial for complete hypersonic delivery vehicles and enough major subsystems to effectively provide access to complete hypersonic missiles. Second, given dual-use concerns, we also recommend a policy of case-by-case export reviews for scramjets and other hypersonic engines and components; fuels for hypersonic use; sensors, navigation, and communication items for hypersonic flight; hypersonic flight controls and design tools and modeling for such uses; and ground simulation and testing for hypersonic systems.

There is at most a decade before hypersonic missiles become militarily significant. This may be just enough time to develop a new international policy. The necessary first step is for the United States, Russia, and China to agree not to export complete hypersonic missiles or their major subsystems. Beyond that, the control list recommended in this report can be the basis for international discussions.

Acknowledgments

The Carnegie Corporation of New York funded this research. Special recognition goes to Carl Robichaud, program officer with International Peace and Security, for his role in launching the project.

In the process of pursuing this research, the RAND team met with some 70 specialists in proliferation, countries and regions, and hypersonic technology. These specialists were from the RAND Corporation itself, the Carnegie Endowment for International Peace, the State Department, Johns Hopkins University Applied Physics Laboratory, the Air Force Research Laboratory, the National Air and Space Intelligence Center, the Defense Technology Security Administration, the Directorate of National Intelligence, the Office of Naval Research, National Security Council staff, the Institute for Defense Analyses, the James Martin Center for Nonproliferation Studies, the National Academy of Sciences panel on hypersonic missile defense, the Department of Commerce, the Army Missile Research Development and Engineering Center, the Arnold Engineering Development Center, and the White House Office of Science and Technology Policy. Our sincere thanks to all the individuals who provided insights into hypersonic technologies and the challenges associated with their development.

We greatly appreciate the assistance of Elizabeth Hammes of RAND's Knowledge Services. She combed through a decade and a half of hundreds of aerospace periodicals to produce most of the references appearing in the discussion of foreign programs.

On October 12, 2016, the RAND team held a workshop with nine of the individuals from the preceding meetings. The workshop discussed the RAND team's interim findings and resulted in extensive revisions to this report. Our thanks go to these nine subject-matter experts.

James Acton of the Carnegie Endowment for International Peace commented insightfully on the RAND team's work in many separate sessions. He deserves particular thanks. The draft was reviewed by Mark Lewis of the Institute for Defense Analyses and Karl Mueller of RAND, who made invaluable suggestions. However, only the authors are responsible for the final product.

Abbreviations

ALV-0	Austral Launch Vehicle
ATLLAS II	Aero-Thermodynamic Loads on Lightweight Advanced Structures
BOS	Background-Oriented Schlieren
cm	centimeters
CNES	National Center for Space Studies (France)
DGA	Defense Procurement Agency (France)
DRDO	Defense Research and Development Organization
FOAS	Future Offensive Air Systems
HCM	hypersonic cruise missile
HGV	hypersonic glide vehicle
HIFiRE	Hypersonic International Flight Research Experimentation
Hikari	High Speed Key Technologies for Future Air Transport Research and Innovation
HSMW	high-speed maneuvering weapon
HSTDV	hypersonic technology demonstrator vehicle
HYCAUSE	Hypersonic Collaborative Australian/U.S. Experiment

HyTEx	hypersonic technology experimental aircraft
IAI	Israel Aerospace Industries
ICBM	intercontinental ballistic missile
IMI	Israel Military Industries
ISRO	Indian Space Research Organization
ITAR	International Traffic in Arms Regulations
IXV	intermediate experimental vehicle
JAXA	Japan Aerospace Exploration Agency
K	kelvins
kg	kilograms
km	kilometers
km/hr	kilometers per hour
LAPCAT II	Long-Term Advanced Propulsion Concepts and Technologies
LEA	flight-test vehicle (Russia)
LFK	Hypersonic Technology Joint Program (Sweden)
m	meters
MaRV	maneuvering reentry vehicle
MJ	megajoule
ms	milliseconds
MTCR	Missile Technology Control Regime
NASA	National Aeronautics and Space Administration
NATO	North Atlantic Treaty Organization

NORAD	North American Aerospace Defense Command
NPT	Nuclear Nonproliferation Treaty
OODA	Observe, Orient, Decide, Act
R&D	research and development
RV	reentry vehicle
SAMP/T	Surface to Air Missile Platform/Terrain
scramjet	supersonic combustion ramjet
SFRJ	solid fuel ramjet
SHEFEX	Sharp Edge Flight Experiment
SHEFEX I	Sharp Edge Flight Experiment I
SHEFEX II	Sharp Edge Flight Experiment II
SHOC	Stand-off High-Speed Option for Counter-Proliferation
SHYFE	Sustained Hypersonic Flight Experiment
UAV	unmanned air vehicle
UK	United Kingdom
WMD	weapon of mass destruction

CHAPTER ONE

Introduction: What This Report Addresses

Under development in a handful of countries, hypersonic missiles are considered game changers for a number of reasons addressed in this report. For example, such missiles are able to defeat most current and envisioned missile defense systems.

This report addresses two key questions:

1. What are the implications of the proliferation of hypersonic missiles to additional nations? That is, why should the United States and the rest of the world be concerned with such proliferation, and why should it be addressed now?
2. What are the possible measures to hinder such proliferation? That is, is it even feasible to hinder the spread of this technology, and who should buy into such an objective and with what measures?

To address these questions, the authors interviewed some 70 specialists in proliferation, countries, and regions. The authors searched through hundreds of articles in aerospace periodicals dating over the last decade and a half. And the authors drew on their own technical and policy experience.

Missiles and other flying vehicles can travel in three speed ranges—subsonic, supersonic, and hypersonic. Subsonic missiles fly at less than the speed at which sound travels (Mach 1), about 1,000 kilometers per hour (km/hr).[1] Supersonic missiles fly above Mach 1. They are generally regarded as flying between Mach 1 and Mach 5, about 1,000 to 5,000 km/hr. Hypersonic missiles, the subject of this report, travel in the high supersonic range at speeds generally regarded as faster than Mach 5, or about 5,000–25,000 km/hr. Put another way, hypersonic missiles fly faster than about one mile to five miles per second.

There are two types of hypersonic missiles currently under development. The first, hypersonic glide vehicles (HGVs), are typically launched by rockets into the upper atmosphere. They are released at altitudes that can vary from around 50 km to higher than 100 km and glide to their targets by skipping along the upper atmosphere. Figure 1.1 illustrates a generic concept of an arrowhead-shaped HGV.

Figure 1.1
Generic Concept of an HGV

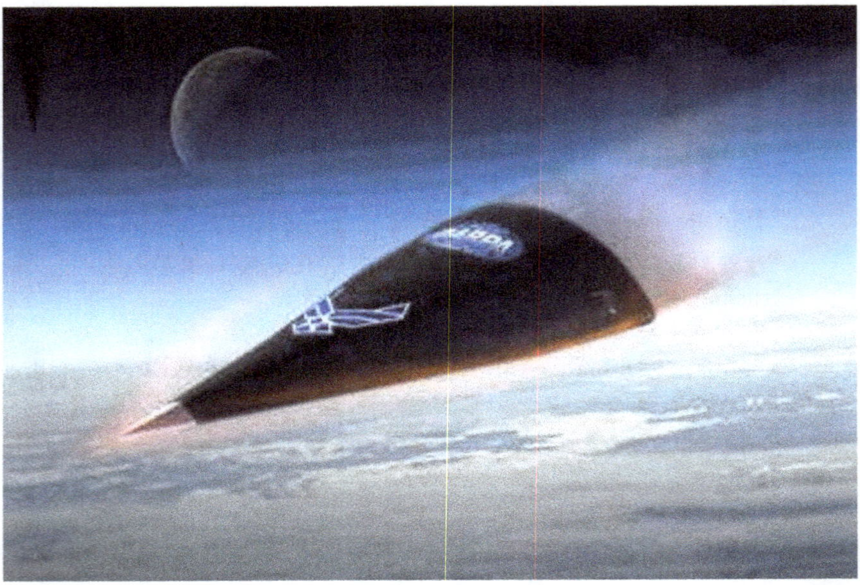

SOURCE: U.S. Air Force illustration.
RAND *RR2137-1.1*

---

1 The speed of sound in the atmosphere varies as discussed in Appendix A; we approximate it here at 1,000 km/hr for ease of discussion.

The second, hypersonic cruise missiles (HCMs), are powered all the way to their targets by rockets or high-speed jet engines. Figure 1.2 illustrates the U.S. WaveRider HCM test.

HGVs differ in important ways from current types of ballistic and cruise missiles. As shown in Figure 1.3, an HGV can vary its impact point and associated trajectory throughout its flight time. HGVs also fly at lower altitudes compared with ballistic missiles. These character-

Figure 1.2
Generic Concept of an HCM

SOURCE: Defense Advanced Research Projects Agency illustration.
RAND *RR2137-1.2*

Figure 1.3
Ballistic Reentry Vehicle (RV) Versus HGV Trajectories

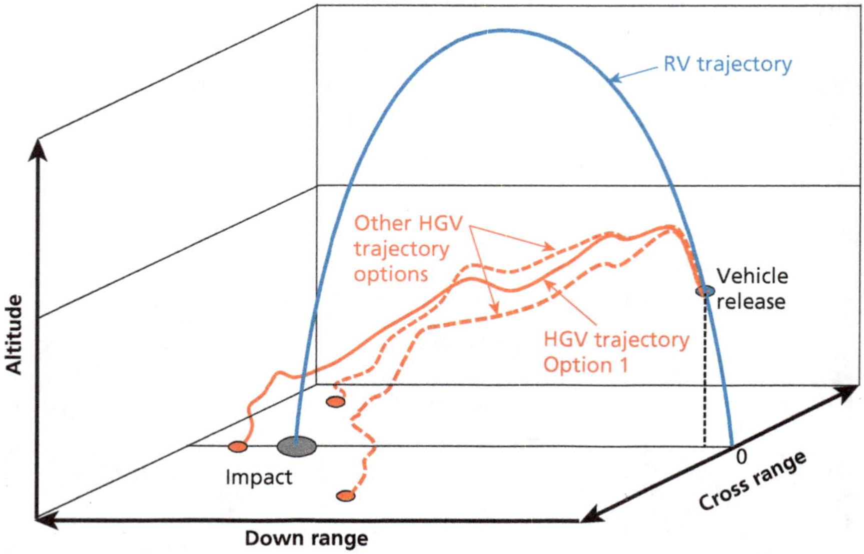

SOURCE: RAND analysis.
RAND *RR2137-1.3*

istics, to be explored in this report, can make hypersonic missiles more threatening and destabilizing than existing missiles.[2]

Perhaps as much as a decade will pass before hypersonic missiles will be ready for military use. By far the leading developers of such missiles are the United States, Russia, and China. Several studies in the public literature have explored the strategic implications of hypersonic missiles (mainly HGVs) in the possession of these three nations, as well

---

2 There are many other potential types of hypersonic weapon systems that could be developed. These include more complex missile systems, manned and unmanned reusable air vehicles, and space launch systems. This report specifically addresses HCMs and HGVs because they are likely the nearest-term. As we discuss late in Chapter Two, these first-generation weapons, but especially HCMs, will provide important flight test platforms to expand the hypersonic flight envelope and to improve hypersonic technologies, in order to develop these more advanced weapon systems.

as possible arms control arrangements among them.[3] This report does not attempt to repeat these studies. Rather, it examines the proliferation of hypersonic missiles and their supporting technologies beyond the United States, Russia, and China.

This report first explores some of the potential strategic implications of the proliferation of hypersonic missile technology beyond the three. It then examines the process of such proliferation. And finally, it discusses possible means for hindering such proliferation. These matters are discussed in the next four chapters, with details in the appendixes.

---

[3] See, for example, James M. Acton, "Silver Bullet? Asking the Right Questions About Conventional Prompt Global Strike," Carnegie Endowment for International Peace, Washington, D.C., 2013; Middlebury Institute of International Studies at Monterey, *Nonproliferation Review,* Vol. 22, No. 2, June 2015. These studies contain numerous references to other literature on the subject.

CHAPTER TWO

Strategic Consequences of Hypersonic Missile Proliferation

To understand the implications of hypersonic missile proliferation, it is necessary to understand the advances these missiles offer compared with current military capabilities. Hypersonic vehicles have been in existence since the dawn of the space age. Manned hypersonic air vehicles were flown over 50 years ago, when the National Aeronautics and Space Administration (NASA) first flew the X-15 hypersonic test vehicle in 1959. (Appendix A contains more details about hypersonic flight vehicles.) The focus of this study, however, is on two new types of hypersonic vehicles and their constituent enabling technologies: HGVs and HCMs.

The principal concerns about HGVs and HCMs are the current development efforts by the major powers (Russia, China, and the United States) and the potential interest by other countries to acquire these systems because of their unique military utility, i.e., their reach and ability to penetrate most air defense systems, derived from the missile's maneuverability, speed, and altitude.[1] It is the combination of these characteristics that makes these systems challenging to develop and to defend against. In contrast, subsonic cruise missiles offer good maneuverability but relatively low speeds, and ballistic missiles offer hypersonic speed but little or no maneuverability.

---

[1] HGVs' capability to maneuver is provided by aerodynamic control surfaces and their flight altitude within the sensible atmosphere (i.e., below 100 km).

We believe that the unpredictable trajectories, resulting in target ambiguity, and the ability to penetrate most defenses, will affect some nations' defense postures and increase instability in some regions. We note that these new missiles will almost exclusively affect nations that are otherwise equipped with effective defenses against ballistic missiles. This may be a substantial number of nations over the coming decades. The next sections describe the major advantages and attributes of HGVs and HCMs and their strategic implications.

Principal Characteristics of HGVs

HGVs are unpowered vehicles that "glide" to their target at the "top" of the atmosphere, reaching between about 40 km to 100 km in altitude. Even in this rarified atmosphere, they are designed to produce lift that is equal to their weight to keep them aloft at hypersonic speeds. A typical operational concept of an HGV involves launching it on a ballistic missile and releasing it at the appropriate altitude, velocity, and flight path angle to enable it to glide to its target. The initial release conditions are driven by the intended trajectory (downrange and crossrange) and the characteristics of the vehicle, e.g., lift and drag. We note that HGV trajectories are very different from maneuvering reentry vehicles (MaRVs) developed in the past. As Figure 2.1 shows, the MaRV trajectory is mostly in ballistic mode above 100 km with some maneuvers executed post-reentry. In contrast, the HGV spends a negligible portion (if any) of its flight in ballistic mode.

The capabilities of hypersonic missiles give them both offensive and defensive advantages. From an offensive perspective, maneuverability can potentially provide HGVs the ability to use in-flight updates to attack a different target than originally planned (within the reach of the weapon system) as shown in Figure 1.3[2] With the ability to fly at unpredictable trajectories, these missiles will hold extremely large areas

---

[2] HGVs are inherently maneuverable from the time they start their glide phase to the target.

Figure 2.1
Typical HGV and MaRV Trajectories

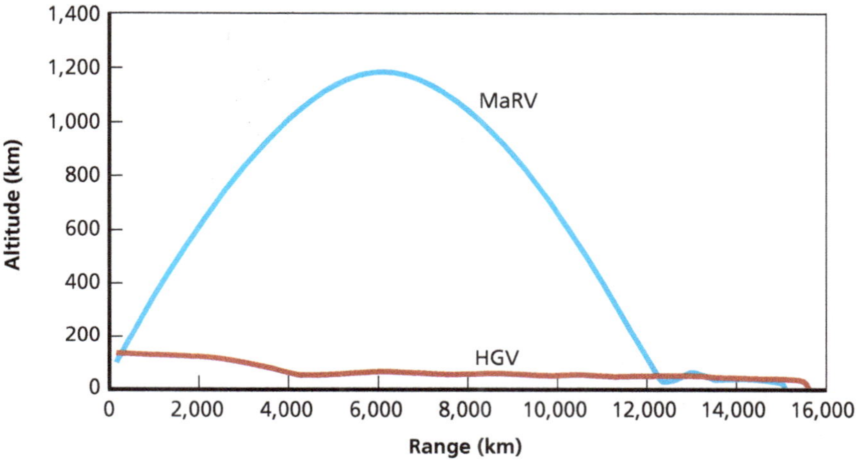

SOURCE: RAND analysis.
RAND *RR2137-2.1*

at risk throughout much of their flights.[3] There are also major defensive differences between MaRVs and HGVs. The post-reentry high–g-force maneuvers for both missiles would challenge terminal defenses, but because the majority of the MaRV trajectory is ballistic, midcourse ballistic missile defense systems that operate in the exo-atmospheric region remain effective against MaRVs but not against HGVs. In other words, a MaRV has all the attributes and vulnerabilities of a ballistic RV with the exception of the post-reentry phase.

Although HGVs are not usually powered, a small propulsion system providing additional velocity or some attitude or directional control could also be integrated into the vehicle. However, the value of such an engine would need to be traded against the costs associated with additional weight and added complexity.

---

[3] Tracking systems cannot estimate a hypersonic vehicle's impact point, which can vary greatly in both downrange and cross-range, until the final phase of flight.

HGVs as Weapons

Defense Penetration

The trajectory and capabilities of HGVs provide them with some unprecedented attributes that may be disruptive to current military doctrines of advanced nations. HGVs have the reach and speed of ballistic missiles, but, unlike these missiles, they fly at lower altitudes and have relatively unpredictable trajectories that can include significant cross-range and terminal maneuvers. These characteristics make HGVs challenging to defend against because they tend to fly outside the altitude and speed envelopes of most modern air and missile defense systems. They can defeat current ballistic missile defense systems because of their unpredictable long-range trajectories, maneuverability, and flight altitudes. Terminal air defense systems would also be challenged by HGVs because of their high speeds and potential endgame maneuverability. Nations that do not possess advanced defense systems capable of defending against ballistic missiles would likely not experience as great a change in threat from these new weapons because they are already vulnerable to ballistic missiles. The possible exception is warning time.

Hypersonic weapons do substantially increase the threat for nations with otherwise effective missile defenses. Hypersonic weapons will not be fielded in quantity for perhaps another decade, and the proliferation to lesser nations would come later—after ballistic missile defenses had been improved and more widely deployed.

Compressed Timelines

Nations that do not possess (or have access to) space-based sensor systems to detect ballistic missile launches and that rely on ground-based sensors, such as radars, to detect incoming mid- to long-range ballistic missiles, could experience a further compression of their decision/response timelines. The reasoning is that typical ballistic missiles tend to fly at higher altitudes than HGVs and should therefore be detectable earlier. Figure 2.2 illustrates this effect. Due to the Earth's curvature and the HGV's low-gliding altitude as compared with that of a similar range ballistic missile, radar or other line-of-site sensors will likely not

Figure 2.2
HGV Versus RV Terrestrial-Based Detection

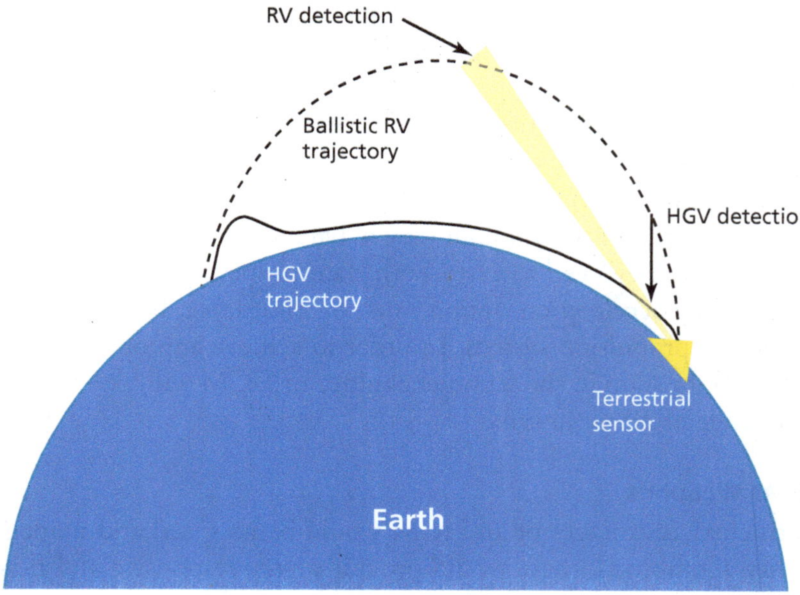

SOURCE: RAND analysis
NOTE: Not to scale.
RAND *RR2137-2.2*

detect an HGV as early as they would a ballistic missile. For example, a radar operating from the surface of a smooth Earth would detect a 3,000-km-range RV about 12 minutes before impact, but would not detect an HGV until about six minutes before impact. We note that potential defensive systems that intend to intercept incoming ballistic missiles before they deploy their payload, e.g., in the boost phase, would retain their effectiveness against HGV weapons.

Principal Characteristics of HCMs

As the name implies, an HCM is a cruise missile that operates at hypersonic speeds. As such, it compresses the defense response timeline and challenges many of the current defense systems because of its high speed and maneuverability. Hypersonic weapons could be launched from the ground, from aircraft, or from ships. An HCM would likely accelerate to around Mach 4 or 5 before an air-breathing engine capable of producing thrust at hypersonic speeds, e.g., a supersonic combustion ramjet (scramjet), further accelerates and then maintains the missile's speed.

There are different options for propelling an HCM to Mach 4 or 5, where the scramjet would take over. Rocket boosters are the most likely option especially for early generation HCMs, because they offer simplicity and affordability, although they may be the largest and heaviest option because they need to carry both their propellant and oxidizer.[4] Of course, any acceleration option must be affordable, because it is a one-time-use propulsion system. In order to achieve appropriate pressures for combustion in the scramjet engine, an HCM will likely cruise at an altitude of 20 to 30 km.

HCMs as Weapons

The principal advantages of an HCM would be its speed and maneuverability. Combined, these would provide a very responsive and flexible offensive weapon that could, for example, hold targets within a 1,000-km radius of the launch aircraft at risk and could strike these targets within several minutes. Cruise missiles are difficult to defend against because of their unpredictable trajectories. The additional speed provided by an HCM, relative to other cruise missiles, would further complicate defense system timelines, as well as potentially be more effective against kinetic defenses, e.g., missile interceptors. Compounding the defensive challenges even further, HCMs would fly at altitudes higher than most current surface-to-air missile systems are capable of reaching. Defenses could be designed to fly higher, but the interceptors still would need to confront the HCM's speed and maneu-

---

[4] There are alternative acceleration systems. For example, a design might employ an expendable jet engine that is capable of providing thrust from a standstill to around Mach 4, at which point the transition to a scramjet occurs. A third option might be a hybrid system that integrates rocket propellant into a ramjet combustor. The rocket would accelerate the missile to low supersonic speeds, followed by ramjet propulsion to around Mach 5 and then an engine flowpath (inlet, combustor, nozzle, etc.) geometry change to enable a transition to scramjet operation.

verability. Furthermore, as described next, an HGV's high kinetic energy affords significant destructive power, even without, or in addition to, the destructive power of an explosive warhead.

Destructive Power from High Speed

Hypersonic weapons can deliver nuclear or conventional warheads. However, another attribute common to both HCMs and HGVs is the potential to use solely kinetic energy to destroy or damage an unhardened target. This is made possible by the combination of their high speed, or kinetic energy, and their accuracy. Their high impact speed can also be leveraged to help defeat underground facilities.[5] Figure 2.3 provides a rough estimate of the effective explosive TNT equivalence of a high-speed mass, such as a conventional strike vehicle with no onboard explosives. The effective TNT equivalence calculation assumed that the explosive force is directional and focused within the approximate cross-sectional area of the impacting vehicle.

Figure 2.3
Destructive Power of a High-Speed Mass as a Function of Speed

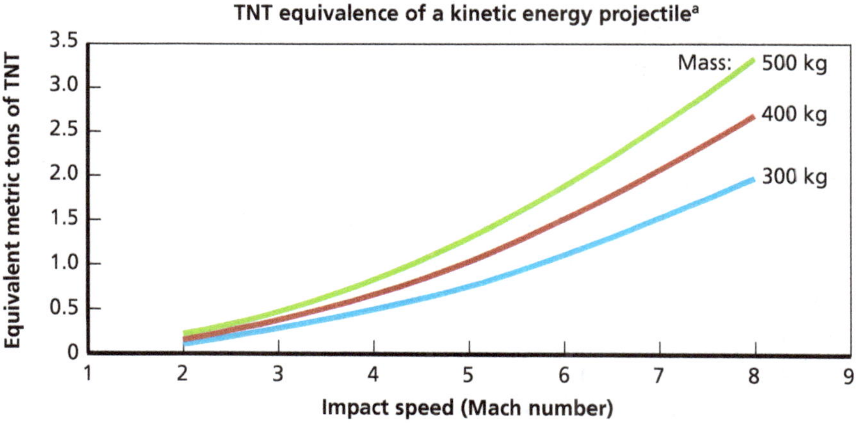

[a] Assumes energy directed and focused along projectile direction and frontal area.
SOURCE: RAND analysis.
RAND *RR2137-2.3*

---

5 However, their penetration capability depends on a combination of speed, weight, shape, and material hardness.

Summary of Challenges for Defensive Systems

As mentioned previously, speed complements hypersonic missiles' maneuverability to significantly increase effectiveness. Defenders with capable terrestrial and space sensors will have only a few minutes to know these missiles are inbound, and lesser adversaries will likely not have any significant warning. Given short timelines and high speed, only very responsive and capable defensive measures would have any chance of defeating the incoming missiles. This likely means that new, space- or terrestrial-based area defense systems, such as boost intercept[6] or highly capable midcourse intercept systems,[7] would be required. These types of systems do not currently exist and would require significant investments to develop and deploy. Advanced terminal (or point) defenses could provide some effectiveness against these high-speed maneuverable missiles. However, such point defenses would likely only be deployed to protect high-value facilities or weapon systems; protecting all potential targets including civilian facilities could be cost-prohibitive. Furthermore, even if a target is equipped with advanced point defenses such that it is able to defend against an HCM or HGV, it may still be vulnerable to salvos of such weapons, especially if these simultaneous attacks use maneuverable vehicles capable of controlling the timing and direction of the attacks.

Defenders may work to develop directed energy defenses, such as lasers, but if such systems were terrestrial-based, they would be challenged by clouds or other atmospheric disturbances and by the need to hit and destroy fast-maneuvering missiles that are equipped with capable thermal protection systems. While a laser beam travels at the speed of light, rendering a near instantaneous time of flight, the beam must dwell continuously and for a significant length of time on a spot on the target to destroy it. The hypersonic weapon's thermal protection system may inherently harden the missile against laser weapons, such that the required laser spot dwell time may be relatively long to burn-

---

[6] Boost intercept occurs during the boost phase of the missile trajectory, i.e., before the payload (RV or HGV) is released.

[7] A midcourse defense system intercepts the payload (RV or HGV) after its boost phase but before its final trajectory phase, i.e., reentry or "dive."

through or sufficiently degrade the thermal protection system (potentially several tens of seconds or longer).[8]

Altitude will also contribute to these missiles' effectiveness, at least in the near term. HCMs will likely be capable of flying at altitudes between 20 km and 30 km, and HGVs will fly at altitudes between about 40 km and 100 km. While the HCM's flight altitudes may be within the upper end of the operating envelope of today's most capable surface-to-air missiles, the combination of altitude, maneuverability, and speed would greatly limit the effectiveness of these defenses. HGVs will fly above the maximum effective altitudes of most surface-to-air missiles, but very likely below the altitudes where exo-atmospheric defenses are designed to intercept inbound RVs.

Long-Term Planning Perspectives for HGV and HCM Technologies

Both HGVs and HCMs offer advanced warfighting capabilities. However, the HCM is also an important stepping-stone to larger manned and unmanned hypersonic vehicles with the potential for military and civilian uses. Prospective applications include military strike and intelligence, surveillance, and reconnaissance aircraft. Furthermore, these vehicles will offer the opportunity to test new flight designs under actual flight conditions. For example, once an HCM is fielded, states will be less reliant on ground test facilities and computer models. Instead, test vehicles will be able to investigate different materials, flight control mechanisms, and flight envelopes under actual flight conditions. Further, availability of flight test data to calibrate ground test facilities and computational models will increase greatly.

---

8 Although we have not calculated required dwell time because of the lack of specific information about HGV-HCM thermal protection systems and about the specific directed energy weapon characteristics, we do know that the thermal protection system is designed to handle very high heat transfers associated with a hypersonic thermal environment. The challenges discussed here are typical of those associated with directed energy weapons.

Strategic Implications of Hypersonic Weapons

Compressed Timelines

The U.S. military uses an acronym to describe the decisionmaking and action process cycle: OODA (Observe, Orient, Decide, Act). These four steps take time, and hypersonic missiles compress available response time to the point that a lesser nation's strategic forces might be disarmed before acting. As an illustration of the time required to act with respect to an existential missile threat, the Nuclear Threat Initiative organization estimated a timeline for a U.S. response to a massive Russian intercontinental ballistic missile (ICBM) attack, as follows:[9]

- 0 minutes—Russia launches missiles
- 1 minute—U.S. satellite detects missiles
- 2 minutes—U.S. radar detects missiles
- 3 minutes—North American Aerospace Defense Command (NORAD) assesses information (2 minutes max)
- 4 minutes—NORAD alerts White House
- 5 minutes—first detonations of submarine-launched ballistic missiles
- 7 minutes—locate president and advisers, assemble them, brief them, get decision (8 minutes max)
- 13 minutes—decision
- 15 minutes—transmit orders to start launch sequence
- 20 minutes—launch officers receive, decode, and authenticate orders
- 23 minutes—complete launch sequence (2 minutes max)
- 25 minutes—Russian ICBM detonations.

This timeline is not, of course, representative of two hostile parties in closer proximity or with less effective warning systems than Russia and the United States. Nor is it representative of less-than-Armageddon possibilities. However, for adjacent enemies within a 1,000-km range, a

[9] Nuclear Threat Initiative, "Is Launch Under Attack Feasible?" web page, August 4, 2016b.

hypersonic missile traveling at ten times the speed of sound could cover that distance and reduce response times to about *six minutes*.[10]

Targets

As discussed earlier, hypersonic missiles increase the threat over current generations of missiles in cases where the target nation has missile defenses. The targets in such nations would primarily be high value and heavily defended. Prime targets could include destroying a nation's leadership and command and control, referred to as "decapitation," to prevent the target nation from responding with an effective follow-on attack. Other key targets could be carrier strike groups, with the objective of striking a key blow or pushing the naval formation further from the coast. And, because of their time sensitivity, strategic forces and storage facilities for weapons of mass destruction (WMDs) could warrant hypersonic attack.

Implications for Targeted Nations

Any government faced with the possibility that hypersonic missiles would be employed against it—particularly in a decapitating attack—would plan countermeasures, many of which could be destabilizing. For example, countermeasures could include devolution of strategic forces' command and control so that lower levels of authority could execute a strategic strike, which would obviously increase the risk of accidental strategic war; or strategic forces could be more widely dispersed—a tactic risking greater exposure to subnational capture. An obvious measure would be a launch-on-warning posture—a hair-trigger tactic that would increase crisis instability. Or the target nation could adopt a policy of preemption during a crisis—guaranteeing highly destructive military action.

To be sure, such measures could be invoked against threats from current types of missiles.[11] But, for nations with effective ballistic mis-

---

[10] This timeline is for illustrative purposes only. We are not suggesting an existential threat from hypersonic missiles in this case.

[11] Pakistan has reportedly taken some of these steps for its tactical nuclear weapons. See Dilip Hiro, "The Most Dangerous Place on Earth," WarIsBoring.com, April 4, 2016.

sile and/or cruise missile defenses in the time frame when hypersonic missiles might proliferate, the hard choices would be forced when facing hypersonic threats.

Advanced nations with adequate resources could take other steps against hypersonic threats. They could strengthen the resilience of their command and control, harden the siting of their strategic forces, and make a deterrent force mobile or sea-based. These tactics may or may not be effective, especially for lesser nations. And they certainly will be expensive—putting them out of reach of some. Even for major powers, the proliferation of hypersonic missiles will create new threats by allowing lesser powers to hold them at risk of effective missile attacks especially against "unhardened" targets, e.g., cities. Over the coming decades, the ability of a lesser nation with a handful of ICBMs to threaten major powers will continue to decrease as wide area missile defenses continue to improve. However, HGVs and HCMs will be more difficult to defend against.

Implications for Major Powers

The ability of hypersonic missiles to penetrate advanced missile defenses will increase the risks for nations with such defenses. Lesser powers with hypersonic weapons may see these weapons as a deterrent against greater power intervention, and feel free to pursue potentially destabilizing regional agendas. Moreover, lesser nations with hypersonic missiles could affect the force deployments of major powers. As noted above, carrier strike groups might be pushed further out to sea or an intervening power's regional military bases might become exposed to more effective attacks.

The Broader Picture of Increased Risk

The ability of hypersonic forces to penetrate defenses and compress decision time could aggravate the instabilities in regions that are already tense—for example, Iran-Israel and North Korea–Japan. Conflicts in these regions could evolve to include major powers aligned on opposite sides. An Israel-Iran conflict, with the United States and much of

Europe aligned with Israel and Russia and perhaps China aligned with Iran, would create new paths for escalation to an even-larger conflict. The basic roles of external actors would not necessarily change—the alignments would stay the same—but external powers might suddenly find themselves in a more-unstable situation in which their patron states are increasingly trigger-happy. As noted previously, lesser powers could gain influence over major powers by threatening a hypersonic attack. At the least, lesser powers might be emboldened if they saw themselves as possessing a deterrent against major power intervention. Finally, because hypersonic weapons increase the expectation of a disarming attack, they lower the threshold for military action.

The powerful capabilities of hypersonic weapons could make the acquisition of hypersonic technology a desirable goal for a number of countries. So, where is there a potential for hypersonic weapons proliferation?

CHAPTER THREE

Ongoing Hypersonic Technology Proliferation

Although the United States, Russia, and China are the furthest along and most aggressively pursuing hypersonic technology, other nations are beginning to build such programs. This chapter describes the current state of research and development (R&D) across more than 20 different countries, based on a sweeping review of aerospace periodical articles dated 2000 through 2016. This chapter focuses on current technological capabilities, past and present R&D programs, a country's projection of its capabilities, wind tunnel facilities and testing ranges, and the rationales for developing hypersonic technology. The details and sources of this research appear in Appendix B.

It is important to note that this chapter does not address the programs of the most-committed and advanced governments—those of the United States, Russia, and China. The progress and capabilities of each of these three countries are already covered extensively in the existing literature. Rather, the purpose of this chapter is to reveal the extent to which more countries have developed programs dedicated to hypersonic technologies. This information will contribute to an assessment of the potential for a nonproliferation effort.

Our research finds that France and India have made the most progress in R&D in hypersonic missile technology, and that these strides have been aided through cooperation with Russia. We also find that pan-European efforts have resulted in several long-term projects dedicated to developing a hypersonic commercial vehicle, aided by Japanese R&D. Australian researchers, by contrast, have principally partnered with U.S. defense entities to develop scramjet technology through a long-running joint program. Outside of these programs, we do not find significant development of hypersonic technology beyond the academic research environment. We outline notable cases of international cooperation and collaborative efforts, followed by an assessment of the problems associated with dual-use hypersonic technologies and the challenges associated with establishing a nonproliferation policy for hypersonic missiles and their constituent technologies.

Committed Governments

After the United States, Russia, and China, the two governments furthest along in the development of hypersonic technology are France and India. While each state is pursuing indigenous capabilities, both have also relied heavily on cooperation with Russia at various stages of development.

France is developing hypersonic cruise missile technology for use in an air-to-surface nuclear weapon delivery vehicle (currently called the ASN4G), but officials suggest that the weapon is still decades away.[1] Other development programs rely upon cooperation with Russia; France planned flight tests of the LEA vehicle (the acronym stands for the Russian phrase for "flight-test vehicle") to be launched from a Russian bomber in Russia between 2014 and 2015 (see Figure 3.1), but it is unclear whether those tests occurred.[2] The vehicle is being developed by the French firms MBDA and ONERA and is still listed as an active program.[3]

India is also working jointly with Russia to develop the BrahMos II hypersonic cruise missile to be used at least in a conventional antiship role (see Figure 3.2). BrahMos II is sometimes said to be an adaption of Russia's Tsirkon hypersonic missile, just as the current Indian-Russian

---

1 "France Studies Nuclear Missile Replacement," *Defense News*, December 1, 2014, p. 22.

2 Michael A. Taverna and Douglas Barrie, "Son of Japhar," *Aviation Week & Space Technology*, Vol. 169, No. 14, October 13, 2008.

3 ONERA, The French Aerospace Lab, "DCPS—System Design and Performance Evaluation: Projects and Research Topics," web page, undated.

Figure 3.1
French LEA

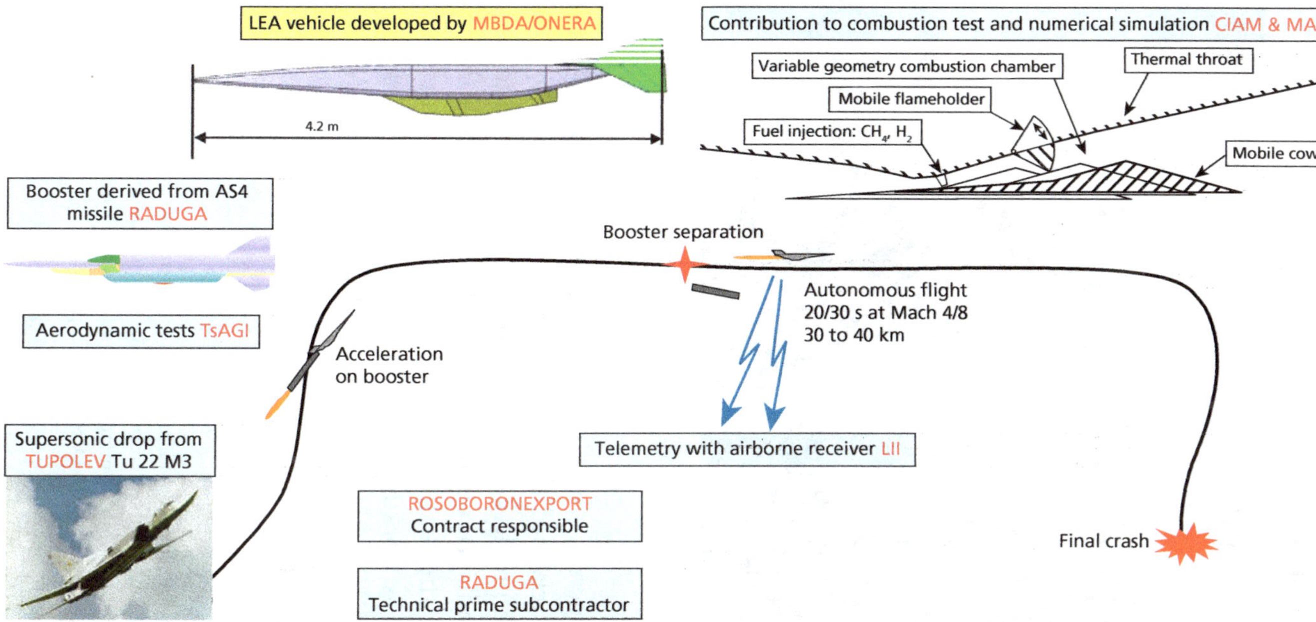

SOURCE: Francois Falempin and Laurent Serre, "French Flight Testing Program LEA Status," Washington, D.C.: NATO Research and Technology Organisation, RTO-EN-AVT-185, undated, p. 17-5, Figure 5.
RAND *RR2137-3.1*

Figure 3.2
Indian-Russian BrahMos II

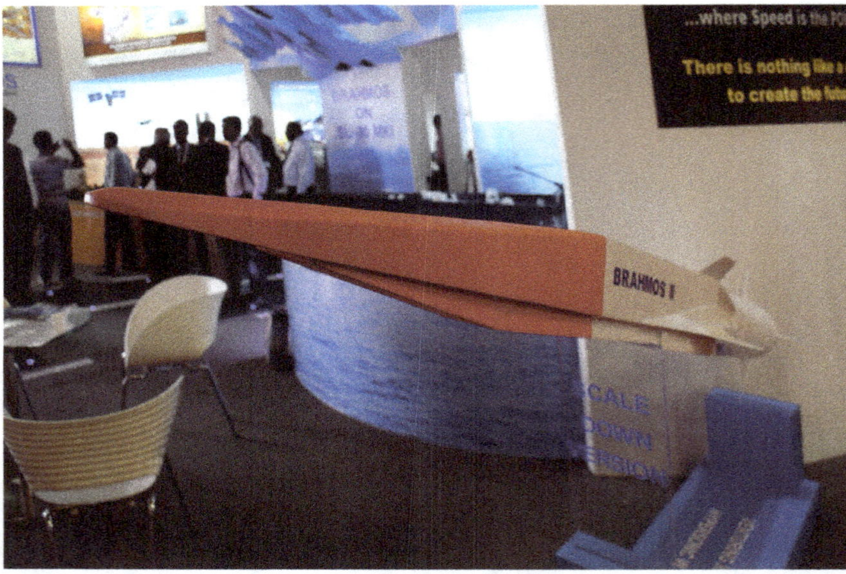

SOURCE: Shiv Aroor via Wikimedia Commons (CC BY-SA 2.5).
RAND *RR2137-3.2*

BrahMos I supersonic missile is an adaptation of Russia's Oniks missile. India has claimed that the BrahMos II would fly by the end of 2017, but such predictions have frequently been revised to later dates. Of concern, India has offered the BrahMos I for export, so the question arises whether the BrahMos II will also be put on the market.[4] Additionally, India is working on an indigenous hypersonic demonstrator vehicle (HSTDV) with the intention of creating an HCM capable of speeds of up to Mach 7. However, the program has consistently failed to meet scheduled milestone goals.[5]

---

[4] Thus far, both Russian and Indian officials have said that they do not intend to export BrahMos II, but it is reasonable to expect that the decision is subject to change. Ulla Uebler, "Analysis and Localisation of Communications Emitters in Strategic and Tactical Scenarios," *Naval Forces*, Vol. 33, No. 5, October 2012, p. 128.

[5] Jay Menon, "Homegrown Hypersonics," *Aviation Week & Space Technology*, Vol. 174, No. 42, November 26, 2012, p. 51.

After France and India, we find three additional governments/entities that are actively pursuing R&D programs in hypersonic technology: Australia, Japan, and the European Union.[6] Similar to the programs being pursued by France and India, each of these programs relies heavily on international cooperation, resulting in diffusion of hypersonic-related technology among these entities.

Australia has a small group of world-class researchers of hypersonics based primarily at the University of Queensland. They have participated in a series of collaborations on scramjet technology with the United States and Europe. The Hypersonic International Flight Research Experimentation (HIFiRE) program is a long-standing collaboration of Australia's Defence Science and Technology Group and the U.S. Air Force Research Laboratory with participation by other Australian and U.S. entities (see Figure 3.3).[7] The program is fairly advanced; in May 2016, researchers launched successful and affordable tests of scramjet prototypes at speeds of up to Mach 7.5.[8] In contrast, Australia's indigenous hypersonic research programs have encountered some problems and setbacks, and, as a result, have seen a reduction in funding over the years.[9]

In 2005, the Japan Aerospace Exploration Agency (JAXA) released a mission statement, JAXA 2025, which detailed the organiza-

---

6 In discussing the "European Union," we refer to activities spanning two or more European Union countries. These include activities of the European Space Agency, government-to-government undertakings (including government-owned-or-controlled aerospace firms), government-to-aerospace-firm (or university) activities, company-to-company (or university) projects, and the activities of single firms with subsidiaries in several countries.

7 Anonymous, "Australia and USA in HiFire Link-Up," *Flight International*, Vol. 170, No. 5063, November 2006, p. 32; "Boeing Announces Involvement in Major," 2007; Yiguang Ju, Skip Williams, and Joanna Austin, "Propellants and Combustion," *Aerospace America*, December 2008, p. 68.

8 Tom Metcalfe, "Blazing-Fast Hypersonic Jet on Track for 2018 Launch," Live Science, May 26, 2016.

9 Guy Norris, "Hyper Hurdles," *Aviation Week & Space Technology*, Vol. 175, No. 38, November 4, 2013; David Lewis and Tom Forbes, "Researchers at University of Queensland Mothball Scramjet Experiment After Failed Test in Norway," *Australia Broadcasting Corporation News*, September 19, 2013; The University of Queensland Centre for Hypersonics, "Current Research Projects" web page, undated-b.

Figure 3.3
Australian–U.S. HIFiRE Scramjet

SOURCE: Australian Hypersonics Initiative at the University of Queensland, Australian Defence Science and Technology Group, and U.S. Air Force Research Laboratory.
RAND *RR2137-3.3*

tion's goal to create a hypersonic commercial aircraft capable of cruising at Mach 5. As a part of this vision, Japan is invested in hypersonic research as a partner in the High Speed Key Technologies for Future Air Transport Research and Innovation (Hikari) program, along with the European Commission and the Japanese Ministry of Economy, Trade, and Industry. Hikari program directors hope to begin experimentation for a future hypersonic vehicle by 2020.[10] Indigenous efforts in Japan focus on a Hypersonic Technology Experimental Aircraft (HyTEx)—another commercial vehicle capable of traveling at speeds of up to Mach 4.5 (see Figure 3.4). This program, however, is still in the early stages of development.[11]

Finally, the European Union has invested in three R&D programs on hypersonic technology: Long-Term Advanced Propulsion Concepts and Technologies (LAPCAT II), Intermediate eXperimental Vehicle (IXV), and Aero-Thermodynamic Loads on Lightweight Advanced

---

10 "JAXA 2025 (JAXA Long-Term Vision)," YouTube, April 9, 2009.

11 Denis Loctier, "Will Hypersonic Passenger Planes Ever be a Reality?" *Euro News*, February 3, 2015.

Figure 3.4
Japanese HyTEx

SOURCE: Promotional photo from JAXA.
RAND *RR2137-3.4*

Structures (ATLLAS II). LAPCAT II is designed to develop a civilian transport airplane capable of cruising at speeds of up to Mach 5 using a hybrid turbo-scramjet engine designed by British defense contractor Reaction Engines (see Figure 3.5).[12] Additionally, the European Space Agency has invested in an experimental suborbital vehicle designed to test atmospheric reentry conditions from (hypersonic) orbital speeds and trajectories, called IXV. In support of these efforts, the ATLLAS II project designs and develops lightweight, high-temperature materials.[13]

---

[12] Hideyuki Taguchi, Akira Murakami, Tetsuya Sato, Takeshi Tsuchiya, "Conceptual Study on Hypersonic Turbojet Experimental Vehicle (HYTEX)," *Transactions of Space Technology Japan*, Vol 7, No. 26, 2009, pp. 27–32.

[13] J. Steelant, M. Dalenbring, M. Kuhn, M. Bouchez, and J. von Wolfersdorf, "Aero-Thermodynamic Loads on Lightweight Advanced Structures II (ATLAS II: Final Report)," *European Space Agency—European Space Research and Technology Centre*, October 2, 2012; J. Steelant, "Sustained Hypersonic Flight in Europe: First Achievements Within LAPCAT II," 17th American Institute of Aeronautics and Astronautics International Space Planes and Hypersonic Systems and Technologies Conference, San Francisco, Calif.: American Institute of Aeronautics

Figure 3.5
European LAPCAT II

SOURCE: Promotional photo from Reaction Engines.
RAND *RR2137-3.5*

And Norway is home to the Andoya Test Center, which provides full-scale hypersonic testing to a host of countries around the world.

R&D in Less-Committed Countries

The RAND research team also reviewed reports of hypersonic research in Brazil, Canada, Iran, Israel, Pakistan, Singapore, South Korea, and Taiwan. These reports describe mainly academic research or proposals by entrepreneurs with low levels of funding, with the exception

---

and Astronautics, Vol. 2243, 2011; Phillip Butterworth-Hayes, "Europe Speeds Up Hypersonics Research," *Aerospace America*, 2008, p. 24.

of Brazil, which is further along in development and testing.[14] While many of those countries have active programs developing supersonic weapons (or have imported such weapons from other countries) or modifying ballistic missile trajectories, we were unable to find evidence of sustained state-sponsored R&D initiatives for hypersonic vehicles. Finally, literature reviews of Belarus, Egypt, North Korea, Poland, South Africa, Turkey, and Ukraine offered little information on what hypersonic research the countries might be conducting, or whether there are any such programs.

International Cooperation

Hypersonic technology can be exported and imported, short-circuiting the slow and costly route of indigenous development. States can share research results, components, testing facilities, test ranges, and other technologies that are critical to the development of hypersonic vehicles. They might do so in an effort to build relationships, increase revenue, or defray some of the costs associated with purely indigenous technological development. We find that each of the three leaders in hypersonic weapons (the United States, Russia, and China) has established cooperative relationships with other states that are seeking to improve their missile technologies. Additionally, international cooperative efforts within the European Union and between European, Japanese, and Israeli researchers suggest that both bilateral and multilateral technology-sharing agreements are growing.[15]

---

[14] João Felipe de Araújo Martos, Israel da Silveira Rêgo, Sergio Nicholas Pachon Laiton, Bruno Coelho Lima, Felipe Jean Costa, and Paulo Gilberto de Paula Toro, "Experimental Investigation of Brazilian 14-X B Hypersonic Scramjet Aerospace Vehicle," *International Journal of Aerospace Engineering*, Vol. 2017, No. 50, May 2, 2017.

[15] Fokker, "Fokker, NLR, Airborne and TU Delft Start Maintenance Centre for Composites," press release, June 16, 2015; F. F. J. Schrijer, B. W. Van Oudheusden, U. Dierksheide, and F. Scarano, "Quantitative Visualization of a Hypersonic Double-Ramp Flow Using PIV and Schlieren," in 12th International Symposium on Flow Visualization, Göttingen, Germany, September 14, 2006.

As discussed previously, Russian cooperation with India has led to significant developments in Indian capabilities. While India remains behind the United States, Russia, and China in their development, close cooperation with Russia has made India a leader among the second tier of states pursuing hypersonic technologies; Russia holds a 49.5 percent stake in BrahMos II.[16] A recent Indian technology-sharing agreement with Belarus (a close Russian ally) may further spread the diffusion of hypersonic technology. We note India's need for significant foreign technical assistance to develop its hypersonic programs.[17] Russian cooperation with France has additionally led to advances as French companies gain access to important testing facilities.[18]

Similarly, the United States has developed a close relationship with Australian researchers through the HIFiRE program—a joint collaboration between the Australian Defence and Technology Group and the U.S. Air Force Research Laboratory. U.S. cooperation with Australia on the HyShot (see Appendix B) and HIFiRE programs over the past 15 years has led to advances in Australian space and defense technologies.[19] Australia is also home to the Woomera Rocket Range, one of the major ranges in the world capable of hosting full-scale hypersonic launches.

Intra-European efforts have produced the LAPCAT II, IXV, and ATLLAS II projects, described above. Despite the United Kingdom's (UK's) vote to exit from the European Union in 2016, to date it does not appear that this will affect UK contributions to the LAPCAT II project. The Japanese Hikari program is also dependent upon European support and technology.[20]

Finally, China recently supplied Pakistan with CM-400AKG high-supersonic (Mach ~ 4) rocket-powered cruise missiles (see

---

16 Uebler, 2012, p. 128.

17 Purohit, Jugal, "Inside the BrahMos Missile Factory," *New Delhi Mail Today in English*, February 20, 2017.

18 Taverna, 2008.

19 Metcalfe, 2016.

20 Loctier, 2015.

Figure 3.6).[21] While one can speculate that this is an attempt to balance the Russian-Indian cooperation on the BrahMos family of missiles, it potentially suggests a future in which supplier states compete in offering hypersonic missiles to their friends and allies.

Claimed Reasons for Pursuing Hypersonic Technology

Many (though not all) of the projects involving international partners claim to be for commercial, nonmilitary purposes. Such peaceful use assertions are frequent problems in nonproliferation policy. Nuclear

Figure 3.6
Chinese Mach 4 Missile Exported to Pakistan

SOURCE: Uncredited image of Pakistan Air Force JF-17 fighter with two mounted CM-400AKGs during flight training.
RAND *RR2137-3.6*

[21] "YJ-12 (CM-302)," *Jane's Air-Launched Weapons*, October 5, 2016.

nonproliferation policy must deal with the issue of "peaceful nuclear explosions"; missile nonproliferation policy must deal with the issue of "space launch vehicles."[22] Both involve hardware and technology that are interchangeable with the lethal items against which the policy is formulated.

Similarly, many hypersonic technology programs may have a dual-use character. Such hardware and technology may eventually be used for space launch and civilian transport of passengers and cargo. However, similar technologies, and in some cases hardware, can contribute to hypersonic missiles. Furthermore, once a nation acquires hypersonic capabilities, its intentions can change. Technology once thought to be of use only to reduce the cost of space launches can be repurposed to create a deterrent effect against regional rivals or to increase the state's prestige in the international community. Ultimately, unless a nation declares outright that it is seeking missile delivery vehicles for its military, there are limits to knowing how the program will end up. This is one of at least five challenges (discussed next) for controlling the proliferation of such capabilities.

Challenges Posed for Controlling Proliferation

While many of the challenges inherent in controlling hypersonic missile proliferation are similar to the problems faced by other nonproliferation regimes, there are a few that stand out as particularly problematic. We identify here five principal challenges that a nonproliferation policy will need to address—challenges that are particularly difficult for controlling hypersonic weapon proliferation.

The first challenge is the widespread nature of hypersonic research among governments, industries, and universities. Some universities and laboratories around the world, from the United States to Israel to Brazil, have wind tunnels capable of testing hypersonic flows. Research on hypersonic fluid dynamics is fairly common, and many major uni-

22 On international controls over peaceful nuclear explosions, see United Nations, Article IV of the Treaty on the Nonproliferation of Nuclear Weapons (NPT), New York, May 2005.

versities have at least one faculty member who teaches and/or conducts research on hypersonic flows. Even without physical test facilities, universities and industries are able to contribute to hypersonic research through computational models and theoretical design. Of course, only a limited number of the activities are cutting-edge. However, given the degree of academic interest in this research, the dissemination of knowledge and research findings on hypersonic technology poses a challenge for any nonproliferation measures.

Similarly, the open research and publication of technological information on hypersonic research generates a unique challenge for a nonproliferation agreement. For example, the American Institute of Aeronautics and Astronautics (AIAA) publishes proceedings from international hypersonic conferences. The AIAA held its Hypersonics 2017 conference at the University of Xiamen in China.[23] In 2014, the Von Karmen Institute in Belgium hosted a lecture series to review the comparative advances of European countries in hypersonic technology. The Von Karmen Institute serves as a testing and educational center for some pan-European hypersonic technology development. This kind of open publication and information exchange makes controlling hypersonic proliferation difficult, posing problems for nonproliferation efforts.

As discussed earlier, problems associated with intent and dual-use also pose significant challenges for a nonproliferation policy. Any policy will be forced to deal with claims that the technology will be only applied to civilian passenger aircraft rather than military applications—no matter the economic questions surrounding such claims, as well as the decades required to bring even an uneconomical civilian system online. The use and proliferation of dual-use technologies can often generate distrust between states and makes controlling hypersonic proliferation particularly difficult.

---

[23] American Institute of Aeronautics and Astronautics, "21st AIAA International Space Planes and Hypersonic Systems and Technology Conference (Hypersonics 2017)," web page, undated-a; American Institute of Aeronautics and Astronautics, "21st International Space Plane and Hypersonic Systems and Technology Conference," web page, undated-b.

Fourth, the nonproliferation measures recommended later in this report do not ban indigenous developments of hypersonic technology. Rather, they seek to control the exports of such technology. This leaves indigenous programs in such states as France and India as potential sources of future exports unless those states agree to export controls.

Finally, although indigenous development faces severe technological barriers (see Appendix C), the prevalence of international cooperation on commercial hypersonic activities can result in the diffusion of information and technologies necessary to the development of hypersonic weapons. This can reduce the costs of future indigenous hypersonic development, accelerating timelines and providing additional routes to export research, components, and/or technologies.

Summary

In addition to the United States, Russia, and China, five countries and/or entities are investing significant amounts of resources into the R&D of hypersonic technologies: India and France are the furthest along, followed by Australia, Japan, and the European Union. It appears that while Russia and the United States have been more willing to develop bilateral agreements for the development of missile systems, European countries and Japan have created joint projects that aim to develop a hypersonic commercial airliner. However, the dual-use nature of hypersonic technology, the widespread nature of hypersonic R&D, open publication of research, and ability of international cooperative ventures to shorten the timelines of indigenous programs all pose significant challenges to nonproliferation measures.

How should concerned parties respond to these challenges?

CHAPTER FOUR

Hindering Hypersonic Missile Proliferation

The growing interest in hypersonic technology and its destabilizing potential if obtained for nefarious purposes present a strong case for exploring options to limit the spread of hypersonic missiles and technology. This chapter examines a number of unilateral and multilateral measures that could be used to prevent or reduce hypersonic missile proliferation or some of its consequences. We conclude by recommending an expanded policy of multilateral export controls.

Unilateral Measures

Currently, United States personnel working on hypersonic missile policy appear to be most concerned about Russian and Chinese developments, not those of other nations. To deal chiefly with technology of possible interest to Russia and China, the United States attempts unilaterally to prevent the spread of hypersonic missiles and some of their consequences by three means:

1. The United States classifies the most sensitive hypersonic technologies. Classification of any technology is generally prescribed through written classification guides, some of which may themselves be classified.
2. The United States restricts the export of some unclassified hypersonic technologies by placing them on export control lists.

For munitions, the International Traffic in Arms Regulations (ITAR) control these lists.[1]

3. The United States is beginning to examine the possibilities for defense against hypersonic missiles. The National Defense Authorization Act for Fiscal Year 2017 requires the Missile Defense Agency to "serve as executive agent for the Department of Defense for the development of a capability to counter hypersonic boost-glide vehicle capabilities and conventional prompt global strike capabilities that may be employed against the United States, the allies of the United States, and the deployed forces of the United States."[2] (Note that defenses against HCMs are not addressed.)

A recent study of such defense possibilities is cautious about their outlook. "HSMWs [high-speed maneuvering weapons] can combine speed and maneuverability between the air and space regimes to produce significant new offensive capability that could pose a complex defensive challenge....At a national strategic level, HSMWs could hold at risk the fundamental U.S. construct of global reach and presence."[3]

Unilateral actions against missile proliferation will have limited effectiveness without reinforcing actions by other key nations and be counterproductive if other major powers do not take similar actions. Russia or China can undercut U.S. restraint. For that reason, it is important to explore possible international measures.

---

1 For examples of Navy restrictions under ITAR, see Lore Anne Ponirakis, "Dense Core Ablative Nosetip Materials for Hypersonic Applications," *Navy Small Business Technology Transfer*, December 17, 2012; Dean Putnam, "Ceramic-Metal Joining for Hypersonic Vehicle and Missile Components," *Navy Small Business Technology Transfer*, January 11, 2016.

2 Public Law 114-328, National Defense Authorization Act for Fiscal Year 2017, Subtitle E, Missile Defense Programs, Section § 1687, December 2016.

3 National Academies of Sciences, Engineering, and Medicine, *A Threat to America's Global Vigilance, Reach, and Power—High-Speed Maneuvering Weapons: Unclassified Summary*, Washington, D.C.: The National Academies Press, 2016.

Multilateral Measures

Negotiations and coordination with other governments take time, so it is worth asking how much time is available for hypersonic missile nonproliferation measures before the hardware and technology are too widespread to contain. It appears that there will be a decade or less during which hypersonic missiles and their enabling technologies will remain in the hands of a few key actors and will not become fielded. Although there are predictions that hypersonic missiles will be ready for military use in the 2017-to-2020 period, the history of such complex systems suggests otherwise. Given the rate at which governments move, now is the time to raise the possibility of the control of such systems with other governments. As the history of other nonproliferation regimes demonstrates, sooner is better than later.

One occasional proposal for controlling hypersonic missiles is to negotiate either a global ban or a nonproliferation treaty to stop their spread. However, the history of technology bans negotiated between the current "haves" and the "have-nots" is not promising. Typically, the have-nots demand a price for their restraint—often in the form of access to civilian forms of the items to be banned. The NPT includes a provision agreeing to share the benefits of "peaceful nuclear explosions," and proposals for a ballistic missile NPT typically include a provision to share space launch vehicle technology.[4] One proposal is to initiate a test ban on hypersonic missiles among the United States, Russia, China, and perhaps France and India.[5] However, all of these proposals for bans run up against the question of whether the United States, Russia, and China—now heavily invested in hypersonic developments—would give up the weapons. Without foreclosing the possibility of bans, this report will look at other options that do not require them.

Another frequent suggestion for dealing with proliferation is to promote confidence-building measures. These measures are designed

---

[4] Richard Speier, "An NPT for Missiles?" in Henry Sokolski, eds., *Fighting Proliferation: New Concerns for the 1990s*, Maxwell Air Force Base, Ala.: Air University Press, 1996.

[5] Mark Gubrud, "Just Say No," *Bulletin of the Atomic Scientists*, June 25, 2015.

to reduce tensions by such means as preannouncement of tests or mutual observation of facilities. However, because they do not necessarily hinder the spread of the hardware and technology in question, their nonproliferation value is questionable.

Yet another approach is to offer incentives to nations to abjure hypersonic missiles. These might be positive incentives such as offers of nonhypersonic military aid in return for hypersonic restraint. However, such an approach raises the classic problem that to pay a price for someone not to do something is to encourage that someone to find more objectionable activities not to do. There are also negative incentives, i.e., sanctions. However, sanctions generally require widespread support, and this requires widespread agreement that the particular instance of the sanctioned activity is sufficiently objectionable—a difficult standard to meet except in the cases of such rogue nations as Iran and North Korea.[6]

Shared defenses against hypersonic missiles are one form of positive incentive that might be considered. The National Defense Authorization Act of Fiscal Year 2017 call for examination of such defenses includes provisions for working jointly with other nations. However, as noted previously, the prospects are not clear for effective defenses against hypersonic missiles. Even shared warning of an impending hypersonic attack—perhaps relying on some form of satellite observation—would, if feasible, offer no more than a few additional minutes of reaction time.

Multilateral export controls are international measures that have already been well tested. These require only the actions of the nations possessing the technology in question, not the have-not nations. As is detailed in Appendix C, hypersonic missile technology is exceedingly complex. For example, igniting a scramjet engine has been compared to lighting a match in a 5,000 km/hr wind. During flight, the shape of a hypersonic missile will change; so flight controls need to be adaptive to compensate for this effect. Propulsion (for HCMs), materials, thermal management, flight control, and testing are challenges even for

---

[6] See Richard Speier, Brian G. Chow, and S. Rae Starr, *Nonproliferation Sanctions*, Santa Monica, Calif.: RAND Corporation, MR-1285-OSD, 2001.

the United States, Russia, and China. Consequently, for other nations, such hypersonic developments could be prohibitively difficult, without experienced foreign support. Because a number of regimes for technology export controls currently exist, there is a substantial body of experience to extend them to hypersonic missiles. We examine such an approach more deeply in the remainder of this chapter.

Potential Export Controls

The United States, Russia, and China are key players in any discussion about the control of hypersonic technology capabilities. No export controls against the spread of such capabilities can be effective unless at least these three nations support them. If one of the three chose to freely export hypersonic weapons, the restraint of the other two would be undercut. Some would add France and India to this group—and with France, its nonproliferation experience might give it an important role.[7]

What would be the attitudes of the three governments toward export controls on hypersonic weapons and their technology? Of course, it is impossible to know this with confidence without approaching them through diplomatic channels to obtain an official response. And such responses can vary from time to time depending on other aspects of the relationships of these governments. The authors met with subject-matter experts on these governments or in some cases officials of the governments. Those meeting with the authors were generally optimistic on the attitudes of the governments toward a nonproliferation policy. Without giving up current programs, the three might very well be disposed to try to prevent further proliferation.[8]

---

7 Open sources leave it unclear as to what limits Russia might be placing on its hypersonic technology cooperation with India. For more details, see Chapter Three.

8 For more on the attitudes of Russia and China, see Middlebury Institute of International Studies at Monterey, 2015. In January 2017, Russia suggested bilateral talks with the United States on hypersonic missiles, but it is not clear whether this would address proliferation aspects; see "Russia's Lavrov Denies Meddling in European Votes, Blasts U.S. Intelligence," *Radio Free Europe Radio Liberty*, January 17, 2017.

The maps in Figures 4.1–4.3 show some reasons why Russia and China might prefer to avoid a world in which hypersonic weapons were widely marketed. Both would face challenges to defend against Japanese hypersonic weapons—Russia at least in its far east and China in its most critical cities and infrastructure. The same Chinese cities and infrastructure would be vulnerable to intermediate-range Indian missiles. To these reasons, one could add the North Atlantic Treaty Organization (NATO) military threats to European Russia; a Poland able to purchase hypersonic missiles on the world market would be especially objectionable to Russia.

The value of a policy shared by the three governments is highlighted when considering the technical barriers to developing hypersonic weapons.[9] The barriers to developing hypersonic missiles are so great that a tripartite embargo on exports of complete hypersonic delivery systems and major subsystems could be effective for several years. And other governments might themselves honor such an embargo as part of a wider effort to ensure that hypersonic missiles are not deployed in their neighborhoods. A simple tripartite embargo, either alone or

Figure 4.1
Illustrative Ranges from Japan

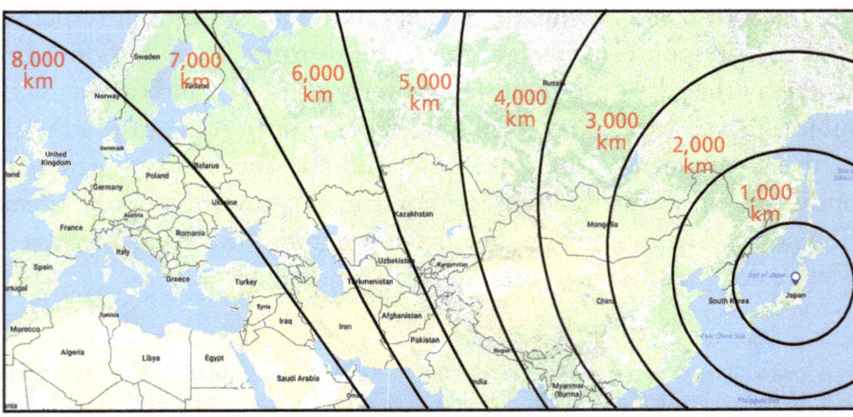

SOURCE: Google Earth with author overlay.
RAND *RR2137-4.1*

9 For technical barriers discussion, see Appendix C.

Figure 4.2
Illustrative Ranges from India

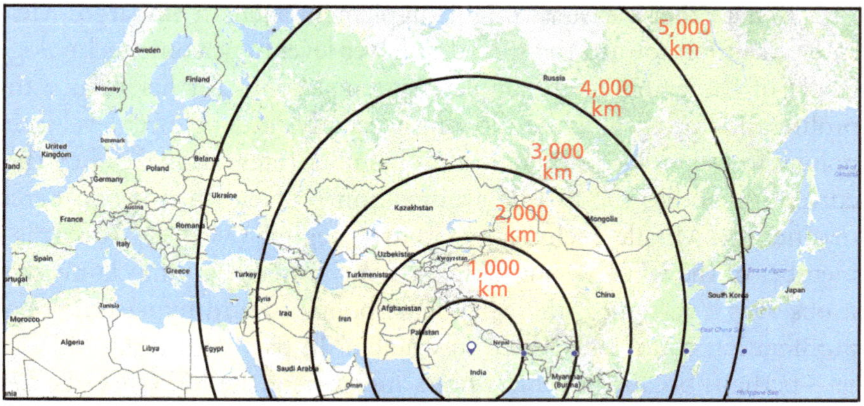

SOURCE: Google Earth with author overlay.
RAND *RR2137-4.2*

Figure 4.3
Illustrative Ranges from Poland

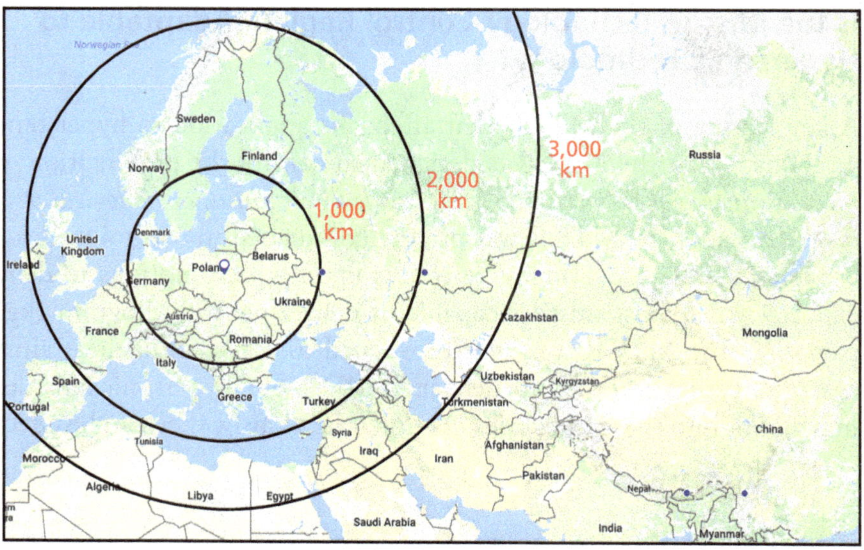

SOURCE: Google Earth with author overlay.
RAND *RR2137-4.3*

with other measures and other supporters, could therefore be the key to hindering hypersonic missile proliferation.[10]

What other measures might supplement such an embargo? Measure of caution toward the spread of lower-level hypersonic technology (short of the embargoed complete systems) could further reduce the proliferation problem while allowing acceptable uses of lower-level technology to be pursued. There is a 35-member international policy that currently handles the missile proliferation problem in this two-tiered manner, the Missile Technology Control Regime (MTCR). Russia is a member of the MTCR, but China is not. However, China claims that it observes a version of the MTCR. A policy toward hypersonic nonproliferation could be adopted in whole or in part within the MTCR, or—perhaps because China is not a member of the MTCR—the key tripartite governments could formulate it separately. Consequently, there are possible arrangements within or outside of the MTCR. By bringing in other nations, the effectiveness of a nonproliferation policy could be substantially enhanced.

Is the Missile Technology Control Regime Adaptable to Hypersonic Technology?

A key feature of the MTCR that affects its application to hypersonic weapons is that the MTCR is designed to control the proliferation of *missiles capable of delivering WMDs (nuclear, chemical, or biological payloads).* Because the MTCR was originally intended to control nuclear-capable missiles, its strongest restraints (strong presumptions to deny exports) are against missiles capable of delivering 500-kilogram (kg) payloads. The MTCR was later broadened to place similar restraints against missiles *intended* to deliver WMDs. But hypersonic missiles may not fit into these categories. As noted previously, they can be effec-

[10] If it were important to recognize the interest in a total ban on hypersonic weapons, the three governments could declare that to be a longer-term objective while implementing the near-term priority of stopping proliferation. However, this report takes no position on the advisability or achievability of such a ban.

tive with a small payload or no payload at all.[11] To redesign the MTCR to direct its strongest restraints against such destabilizing missiles would be a major change in the MTCR's focus—but not an impossible one. Consequently, it will be worth exploring whether it is feasible to place all hypersonic controls in the MTCR or whether to look at other solutions.

Other possibilities would be to ensure that the lesser restraints of the MTCR (case-by-case export application reviews) cover hypersonic hardware and technology. These lesser restraints can be effective. The MTCR includes extensive information exchanges and a "no-undercut" rule (see Appendix D) that help to coordinate the restraint of 35 governments.

Another option might be a hybrid approach with (1) the United States, Russia, and China declaring strong restraints against the export of complete delivery systems and their major subsystems, and (2) the MTCR requiring case-by-case export reviews of lesser components. In fact, the MTCR already requires such reviews of items like scramjets and their (currently undefined) components, so it would not be a stretch to cover other hypersonic items similarly.

Whatever approach is to be taken, it is likely that the final policy would, like the MTCR, strongly hinder the export of some items and allow the export of others. The MTCR strongly hinders the export of rockets and unmanned air vehicles capable of delivering a 500-kg payload to a range of 300 km. It also strongly hinders the export of any missiles intended to deliver WMDs. However, it allows several classes of activities and, in some cases, does not affect them at all. Such allowed activities include the export of manned aircraft, the tightly controlled export of 500-kg/300-km–capable systems on a rare basis, the indigenous development of missile systems, the export of lesser components on a case-by-case basis after examining the end-use and the end-user,

[11] In Appendix D, the issue is raised whether the MTCR controls an HGV as an RV. The MTCR applies its strongest restraints to RVs usable in missiles of specific capabilities.

and the sharing of benefits without the sharing of hardware (e.g., the provision of space launch services without the export of rockets).[12]

At this point, it is appropriate to note the potentially important role of France in a hypersonic missile nonproliferation policy. France is the point of contact in the MTCR, the central point handling documents and hosting intercessional meetings that explore new issues. Moreover, France is perhaps the leading developer of hypersonic technology after the United States, Russia, and China. Whether or not France participates in initial policy actions by the primary three governments, it could be central in coordinating the expansion of any policy to a wider set of international participants.

Recommended Items to Control

This report recommends items that should be subject to new export restraints. The details are laid out in Appendix D. But how should one implement such restraints?

The basic requirement is that the United States, Russia, and China agree on export restraints that they will not undercut. Without such a tripartite sponsorship, any policy will be exceedingly weak. The minimum tripartite agreement would need to embargo complete hypersonic missiles and their major subsystems. As is described in detail in Appendix C, without complete missiles, most potential proliferators would face a long and difficult process to obtain such weapons.

Once a basic tripartite agreement is reached (or in parallel to it), a higher number of nations can agree on a broader set of export restraints. As noted above, we believe that France could play a central role in this process. We do not need to prejudge whether this process would take place within or outside of the MTCR, but the MTCR is well suited for much of the effort.

---

12 The MTCR website provides details on the MTCR Guidelines, which set out the policy rules, and the MTCR Annex, which lists the items controlled by the policy. See Mission Technology Control Regime, "MTCR Guidelines," web page, undated-b; Mission Technology Control Regime, "MTCR Annex," web page, undated-c.

A strong presumption of export denial should be imposed on three items: (1) complete HGVs, (2) complete HCMs, and (3) warheads for HGVs and HCMs.[13]

Case-by-case export reviews should be required for (1) scramjet and other hypersonic engines and their components, (2) fuels for hypersonic use, (3) materials and thermal protection hardware for hypersonic flight, (4) sensors, navigation, and communication items for hypersonic flight, (5) hypersonic flight controls, (6) design tools and modeling for such uses, and (7) ground simulation and testing for hypersonic systems. Details of such controls appear in Appendix D.

Such a two-tier control system would allow some international cooperation on civilian uses of hypersonic technology. However, the authors of this report are skeptical of the optimism about such systems as hypersonic airliners. As is discussed in Appendix C, the economics of hypersonic airliners is dubious, as is the long-term resolve of governments to spend billions of dollars on a project that will take decades for an uncertain outcome. Claims for civilian uses should be reviewed with caution.

Appendix D uses the MTCR format to give examples of how such items might be defined and how they might fit into the existing MTCR Annex.

---

[13] See Appendix D for further details and definition of complete delivery vehicles.

CHAPTER FIVE

Conclusions

The world would be safer if the proliferation of hypersonic missiles was strongly hindered. Such missiles are a new class of threat because they are capable both of maneuvering and of flying faster than 5,000 km/hr. These features enable such missiles to penetrate most missile defenses and to further compress the timelines for response by a nation under attack. The proliferation of such missiles beyond the United States, Russia, and China could result in other powers compressing their response timelines in ways that set their strategic forces on hair-trigger states of readiness—such as a strategy of "launch on warning." And such proliferation could enable such states to more credibly threaten attacks on major powers.

The diffusion of hypersonic technology is under way in Europe, Japan, Australia, and India—with other nations beginning to explore such technology. Proliferation could cross multiple borders if hypersonic technology is offered on world markets.

There is probably less than a decade available to substantially hinder the potential proliferation of hypersonic missiles and associated technologies. The unavoidable requirement is for the United States, Russia, and China to agree on a nonproliferation policy. A relatively simple and effective first step would be for these three governments to embargo complete hypersonic delivery vehicles and their major subsystems. Beyond that, there are various possibilities for placing controls on a wider range of hardware and technology. France could play a key role in bringing other governments into agreement on a broader control policy. The MTCR could provide a mechanism for implementing such a policy or, at least, could serve as a model for an appropriate approach.

There is reason to be optimistic about the potential effectiveness of hypersonic missile export controls. There appears to be interest in hypersonic missile nonproliferation and at least a few years available for relevant governments to put a policy in place. The technical and economic barriers to developing hypersonic technology are great enough to add to the effectiveness of a nonproliferation policy.

The key is time. Governments move slowly, and hypersonic technology development is gradually spreading and becoming embedded in government programs. Nonproliferation discussions should begin while there is still time.

APPENDIX A

The Hypersonic Flight Regime

Introduction

By convention, hypersonic speed is reached when the Mach number exceeds five (M > 5). An object traveling slower than the speed of sound of its surroundings, i.e., typically air, is said to be in the subsonic regime. Large modern airliners travel at the upper end of the subsonic regime. An object traveling faster than the speed of sound, but less than Mach 5, is said to be moving supersonically. The speed of sound in a gas medium, e.g., air, is proportional to the square root of the gas temperature, as follows:

$$a \propto \sqrt{T_{air}}, \quad \text{(Equation 1)}$$

where

$a \sim$ speed of sound
$T_{air} \sim$ the local air temperature
$\alpha \sim$ proportional to.

Mach number is then,

$$M = V/a,$$

where V is the speed of the vehicle.[1]

---

[1] The Mach number is a dimensionless value defined as the ratio between the object speed and the local surrounding, e.g., local atmosphere.

Man-made vehicles operating in the hypersonic regime have been flying for more than 50 years. NASA first flew the X-15 hypersonic test vehicle (shown in Figure A.1) in 1959. The X-15 was a hypersonic, rocket-powered aircraft. In 1967, it set an unofficial world record by flying at an altitude of over 100 km at a speed equivalent to a Mach number of 6.7 (or 6.7 times the local speed of sound). There have been other man-made vehicles operating in the hypersonic regime, such as reentry capsules, e.g., Apollo and Soyuz, as well as reusable launch vehicles, e.g., the Space Shuttle. Additionally, RVs used on ICBMs also reenter and travel through the atmosphere at hypersonic speeds. Satellites orbit at speeds similar to those attained by RVs. However, given that satellites operate in the near vacuum of space, and sound does not travel in a vacuum, the Mach number is not defined and is not a meaningful parameter for vacuum conditions.

These different hypersonic vehicles mentioned experience different heating environments that drive the design of their thermal protection systems. Satellites operate in near-vacuum conditions and therefore do not experience the intense heating rates and pressure loads caused by the atmosphere. Reentry capsules and reusable launch vehicles are subjected to high heating rates and pressures resulting from flying through

Figure A.1
X-15 Hypersonic Test Vehicle

SOURCE: NASA photo.
RAND *RR2137-A.1*

the atmosphere at high speeds. However, these vehicles' large dimensions, specifically their large nose radius, reduce their heating rates. Additionally, their trajectories can be designed to minimize the total heat transfer induced by the high-speed flow over the vehicle body.[2] The total heat transfer on the body can be roughly estimated from:

$$Q_{total} \ \alpha \ \int (\rho/Rnose)^{0.5} \, v^3 \, dt \,, \qquad \text{(Equation 2)}$$

where

Q_{total} ~ measure of total time-integrated heat transfer
α ~ proportional to
ρ ~ local air density
v ~ velocity magnitude (speed)
Rnose ~ nose radius of the vehicle
t ~ time.

In other words, the four main parameters influencing the total heat transfer are the vehicle dimension, speed, density, and total flight time. As Equation 2 indicates, the larger the nose radius, the smaller the heat transfer on the nose of the vehicle. Trajectory shaping, i.e., velocity and altitude, can also be used to manage the total heat transfer on an RV while meeting other input requirements and constraints, e.g., range, maximum deceleration, and time of flight. Hypersonic weapons have different constraints and requirements compared with reentry bodies. HGVs and HCMs will tend to have sharp leading edges, i.e., a small nose radius, which will increase the heat transfer, as indicated by Equation 2. The mission of the HGV or HCM is to travel long ranges at high speeds; therefore, two of the major parameters in the total heat equation, i.e., velocity and time, cannot generally be reduced. The remaining trajectory parameter that can be somewhat varied is the density, which is a function of flight altitude. However, the lift needed to maintain the vehicle flying is given by:

---

2 The *heat transfer* is a measure of heat energy applied to a surface per unit of time per unit of area.

$$L \propto C_L \rho v^2, \qquad \text{(Equation 3)}$$

where

$L \sim$ total lift
$C_L \sim$ lift coefficient.

For a cruising HCM or gliding HGV, the lift must be equal to the weight, and slightly higher if the vehicle is maneuvering. (To be sure, the lift on the body may be modified by a centrifugal force component in the direction of the lift as the vehicle approaches orbital velocity.) Therefore, for a given vehicle design and velocity, a minimum density (or maximum altitude) exists to maintain the needed vehicle lift. As the velocity decreases, the density must increase to maintain the same lift, i.e., the altitude decreases.

RVs used in ballistic missiles are also different in that they do not produce any significant, continuous lift and, in general, have a higher peak of heating rate than that of an HGV. However, the total heat transfer will be significantly smaller because of its shorter aerodynamic heating, or heat soak, time. Another major difference between an RV and an HGV or HCM is the fraction of time during flight that the vehicle spends within the atmosphere. An ICBM RV spends most of its flight time, about 80 percent or more (depending on range and reentry angle), outside the atmosphere. On the other hand, an HGV spends more than 80 percent (in some cases 100 percent) of its flight time within the sensible atmosphere, i.e., below 100 km in altitude. And, of course, HCMs spend 100 percent of their time in the atmosphere.[3]

In summary, total heating transfers on HGVs and, in some cases, HCMs significantly exceed those on previous vehicles.

---

[3] For general references on hypersonic technology, see John David Anderson, *Hypersonic and High Temperature Gas Dynamics,* Washington, D.C.: American Institute of Aeronautics and Astronautics, 2000; John J. Bertin, *Hypersonic Aerothermodynamics*, Washington, D.C.: American Institute of Aeronautics and Astronautics, 1994; and W. H. Heiser and D. Pratt, *Hypersonic Airbreathing Propulsion*, Washington, D.C.: American Institute of Aeronautics and Astronautics, 1994, p. 100.

APPENDIX B

Survey of Foreign Hypersonic Activity

As discussed in Chapter Three, this is a review of hypersonic programs in selected nations. It is based on a survey of aerospace articles dating from 2000 through 2016 and a few more recent articles. It includes select supersonics programs because they can be stepping stones to hypersonic development. It also includes enumerations of major facilities because they demonstrate a commitment to R&D.[1] We begin with a description of pan-European efforts that will inform some discussion throughout the rest of the appendix, and then we proceed alphabetically by country.

This appendix focuses only on a country's most advanced technology. As a result, some advanced missile or aircraft systems may be passed over in discussing countries with more developed hypersonic programs. By contrast, in countries with little investment in hypersonic R&D, the appendix may focus more heavily on supersonic and even subsonic systems. The appendix is not comprehensive. The aim is to inform the reader about trends and the scope of hypersonic developments.

---

1 In cases where the aerospace trade press has covered a development, this appendix's footnotes give representative samples of the reporting. More coverage can readily be found on the Internet. The specifications described in this appendix are summary descriptions of the major capabilities of wind tunnel facilities around the world that we believe will be helpful to the non-technical reader. We recognize that these are not complete descriptions of the full capabilities of each facility and may omit some details important to the technical community. For a fuller depiction of facility capabilities, please see the articles referred to in the footnotes.

We note that the contents of this appendix are solely based on an open literature search and have not been verified with any representative from the subject countries. All information is up to date as of May 17, 2017.

European Union

As noted in Chapter Three, in discussing the "European Union," we refer to activities spanning two or more European Union countries. Because such activities are multinational, we discuss them here before discussing individual country programs.

The European Union has been involved in funding and developing several different initiatives that advance research and production of high supersonic (e.g., the Meteor Missile) and hypersonic technology (e.g., the LAPCAT II, ATLLAS II, and IXV). While the Meteor Missile program is designed as a defense project, other (hypersonic) programs appear to be aimed at civilian transport systems and space RVs.

Advanced Supersonic Technology Initiatives

The European Union is currently developing and finalizing production on the Meteor Missile, a ramjet-powered air-to-air missile capable of traveling at speeds of up to Mach 4 with a range of more than 100 km.[2] The ramjet was produced by the German contractor Bayern-Chemie. Funding for the missile is jointly shared between six European nations: the United Kingdom (39.6 percent), Germany (16 percent), France (12.4 percent), Italy (12 percent), Sweden (10 percent), and Spain (10 percent).[3] It entered into service with the Swedish Air Force in 2016 on the Saab Gripen strike fighter, and is expected to be operational on

2 IHS Jane's 360: News and Defense Headlines, "Air Launched Weapons," web page, July 30, 2006.

3 "European METEOR Missile Test Fired over Sweden," *Defense Update*, undated.

the Eurofighter Typhoon in 2018.[4] So far, the Meteor Missile has been exported to Qatar, Egypt, India, and Saudi Arabia.[5]

Hypersonic R&D

The European Union has primarily invested in three R&D programs using hypersonic technology: LAPCAT II, ATLLAS II, and IXV. Each of these projects focuses on different elements of research and design for hypersonic technology.

The LAPCAT II project is intended to develop a civilian transport airplane. The project envisions using a hybrid turbo-scramjet engine designed by the British defense contractor Reaction Engines to reach speeds of Mach 5 to Mach 8. This $10 million program began in October 2008 as a four-year investment but is still active as of 2016.[6] Partners on this project also involve the European Space Agency, DLR (Germany), CIRA (Italy), MBDA (UK), ONERA (France), the University of Rome, and the University of Brussels.[7]

Additionally, the European Space Agency has been pursuing research for a hypersonic vehicle that travels at Mach 5–6. Designated the ATLLAS II project, the project focused on designing and developing lightweight, high-temperature materials capable of withstanding high-speed travel.[8] The total cost of the initiative (begun in 2011 and expected to last four years) was $6.5 million, coordinated by the European Space Agency and the European Space Research and Technology Center.[9] Partners in the project include DLR, MBDA-France, ONERA, Sweden's FOI, Italy's Alta SPA, the UK's Gas Dynamics Limited, and other university partners across the European Union. The final report, completed at the end of 2015, details a design for high-

---

4 Craig Hoyle, "PARIS: MBDA on Target for Meteor Introduction," *Flight Daily News*, June 16, 2015.

5 Hoyle, 2015.

6 J. Steelant, "Hypersonic Technology Developments with EU Co-Funded Projects," *Defense Technical Information Center*, September 2010.

7 Steelant, 2010.

8 Steelant, 2011; Steelant et al., 2012.

9 Butterworth-Hayes, 2008, p. 24.

speed flight (along with a feasibility study) that optimizes an aerodynamic, propulsive, structural, and thermal layout.[10]

Finally, in February 2015, the European Space Research and Technology Center launched the IXV, an experimental suborbital RV designed to test atmospheric reentry conditions from (hypersonic) orbital speeds and trajectories.[11] The vehicle is designed to reach low-Earth orbit heights, but never makes a full rotation around the Earth. It is intended to be a reusable satellite launch vehicle that is able to reenter the Earth's atmosphere after reaching a maximum altitude of 256 miles.[12]

Australia

The Australian government has sponsored several projects and collaborations with U.S. agencies in the field of hypersonics. Australia's Defence Science and Technology Group has active collaborations with the U.S. Air Force Research Laboratory, the University of Queensland, and Boeing, among others. The Royal Australian Air Force operates one of the world's premier research and testing centers at Woomera Test Range in South Australia. Researchers at the University of Queensland have also had active collaboration with individual research groups from France, Germany, Belgium, the UK, Japan, India, and China.

Current Advanced Supersonic Technology

Australia currently does not have any investments in ramjet-powered missile or defense acquisitions. While it possesses supersonic missiles (such as the AIM-120 advanced medium-range air-to-air missile), the literature suggests that the Australian military has not procured ram-

---

10 Steelant et al., 2012.

11 Gareth Jennings, "Meteor Trials Near Conclusion," *Jane's Missiles & Rockets*, July 14, 2011.

12 Karl Tate, "How Europe's IXV Space Plane Works (Infographic)," *Space.com*, February, 9, 2015.

jet-powered projectiles or missiles, nor invested any R&D efforts to develop these capabilities.

Hypersonic R&D

The Australian HyShot program, which began testing in 2001, led to extensive collaboration with the United States. The Hypersonic Collaborative Australian/U.S. Experiment (HYCAUSE) was a joint venture by the Defense Advanced Research Projects Agency and U.S. and Australian universities to develop technology for air-breathing scramjet engines, with tests beginning in 2007.[13] In January 2007, Boeing contributed $2 million to create the Boeing-Australia Hypersonics Research Project, which was intended to be a three-flight demonstration of a scramjet engine, but this project was later merged with the ongoing HIFiRE program.[14]

Australia is heavily invested in the HIFiRE project—a six-year (initially, then extended), more than $54 million partnership between Australia and the United States.[15] Building on the success of the HyShot program, HIFiRE was jointly established by the U.S. Air Force Research Laboratory and the Australian Defence Science and Technology Organization (which later became the Defence Science and Technology Group) to "investigate the fundamental science of hypersonics technology and its potential for next generation aeronautical systems."[16] It has launched a series of successful tests, the most recent of which (May 2016) reached its target speed of Mach 7.5 at an altitude of 173 miles.[17] Researchers say that the program is on target

---

[13] The University of Queensland Centre for Hypersonics, "About HyShot Program," web page, undated-a.

[14] "Boeing Announces Involvement in Major Australia-U.S. Hypersonics Research Project," *Defense Daily International*, Vol. 8, No. 2, January 12, 2007, p. 1.

[15] Anonymous, 2006; "Boeing Announces Involvement in Major Australia-U.S. Hypersonics Research Project," 2007; Yiguang Ju, Skip Williams, and Joanna Austin, "Propellants and Combustion," *Aerospace America*, December 2008, p. 68.

[16] The University of Queensland Centre for Hypersonics, undated-a.

[17] Metcalfe, 2016.

to test fly a commercial hypersonic scramjet engine up to Mach 20 in 2018.[18]

Australia's indigenous space program that uses hypersonic technology is called "ScramSpace." It has resulted in successful ground tests up to Mach 14 and flight tests up to Mach 8 focused on the free-flying scramjet.[19] However, the program has run into problems in the past few years, with a 2013 test ending in failure when a nozzle disintegrated on launch. As a result of these hurdles and other government priorities, funding for the project has dwindled to just $5 million.[20]

In August 2015, the University of Queensland partnered with Heliq Advanced Engineering to launch the Austral Launch Vehicle (ALV-0), which is a three-stage space launch system with a reusable first stage.[21] A reusable rocket booster propels the vehicle initially, then once the vehicle reaches speeds of Mach 5, scramjets are expected to take over for the second stage and fly at speeds of up to Mach 10 (the rocket then flies back to base using wings and propellers). [22] With this design, the innovation allows the "hypersonic community to join the space community."[23] This three-stage space project, called "SPARTAN," would use hypersonic technology to develop a satellite launching system that is 95 percent reusable.[24] A simple rocket rather than

---

[18] Metcalfe, 2016.

[19] Guy Norris, "Australia Pushes Toward Space with Hypersonic Effort," *Aviation Week & Space Technology*, Vol. 173, No. 15, April 25, 2011.

[20] Guy Norris, "Hyper Hurdles," *Aviation Week & Space Technology*, Vol. 175, No. 38, November 4, 2013; David Lewis and Tom Forbes, "Researchers at University of Queensland Mothball Scramjet Experiment After Failed Test in Norway," *Australia Broadcasting Corporation News*, September 19, 2013; The University of Queensland Centre for Hypersonics, "Current Research Projects" web page, undated-b.

[21] Darren Quick, "Scramjet-Based Project Looks to Blast Australia into Space 2015," *New Atlas*, August 10, 2015.

[22] Norris, 2015; Quick, 2015.

[23] Quick, 2015.

[24] Norris, 2015, Quick, 2015; UQ News, "Launching Australia into Space," The University of Queensland Centre for Hypersonics, August 10, 2015.

the hypersonic air-breathing accelerator propels another version of the vehicle.[25]

Research and Testing Facilities

The Royal Australian Air Force owns and operates one of the world's premier hypersonic testing facilities at the Woomera Test Range in South Australia. The facility is the largest land-based weapons test facility in the world. In June 2016, Raytheon was awarded $297 million to upgrade its capabilities and prepare for performance tests of the F-35 Joint Strike Fighter.[26]

Australia is home to seven hypersonic wind tunnels used by both private researchers and government programs. The University of Queensland Centre for Hypersonics operates five of the seven, with tunnels able to test speeds of 0.29 Mach to almost Mach 30 (10 km per second).[27] The other two facilities are located at the University of New South Wales and the Royal Melbourne Institute of Technology School of Science, Engineering, and Technology.

University of Queensland

1. *Drummond Tube/Tunnel*: Four tubes that test speeds of up to Mach 4. Conical nozzles can also test up to Mach 7 with temperatures up to 3,000 kelvins (K).
2. *T4 Free Piston Driven Shock Tunnel*: Tests speeds of up to Mach 10. It has been used to test such high-speed RVs as the U.S. Space Shuttle and Japan's HYFLEX vehicle.
3. *X1 Free Piston Driven Expansion Tube*: Four tubes that test speeds of up to Mach 4.7 at temperatures up to 15,900 K.

---

[25] Quick, 2015; Guy Norris, "Subscale Reusable Launch System Demonstrator to Fly This Year," *Aviation Week & Space Technology*, July 22, 2015.

[26] Raytheon, "Raytheon Tech Forges the Future of the World's Largest Test and Evaluation Range," web page, August 8, 2016.

[27] Malinda Goodrich, Jenele Gorham, Wm. Noel Ivey, Sarah Kim, Marieke Lewis, and Carl Minkus, "Wind Tunnels of the Eastern Hemisphere," Washington, D.C.: *The Library of Congress*, August 2008.

4. *X2 Super Orbital Expansion Tube*: Four tubes that test speeds of up to Mach 4.8 at temperatures up to 11,500 K.
5. *X3 Free Piston Driven Expansion Tube*: Large super-orbital expansion tube capable of testing speeds of up to 10 km per second (approximately Mach 30).

Royal Melbourne Institute of Technology

6. *Amrad High Speed Teaching Wind Tunnel*: Little information is available on this tunnel.

University of New South Wales

7. *T-ADFA Shock Tunnel*: Tests speeds of Mach 8 to Mach 11, at total temperatures up to 6,000 K. The test section size measures 0.65 meters (m) in diameter. The Australian Defence Force Academic School of Aerospace operates it.[28]

Belgium

As the principal seat of power for the European Union, Brussels would conceivably benefit from commercial hypersonic ventures—most imagined flight routes for a hypersonic vehicle include Brussels as a primary terminal (however, many other countries might also benefit from a hypersonic airliner if it can be made commercially viable). Dozens of organizations are involved in hypersonic technology development inside the European Union, including many that are attempting to develop a commercial airliner that can fly from Brussels to Sydney, Australia, in less than four hours. Contributors to this program include: the Von Karman Institute, the European Space Agency, the European Space Research and Technology Centre (ESTEC), the German Aerospace Centre (DLR), Reaction Engines, French defense contractor ONERA, the Italian Aerospace Research Centre (CIRA), Cenaero, Snecma, Airbus Group (formerly known as EADS), defense

---

[28] Goodrich et al., 2008.

contractor MBDA, Gedvel Defense Logistics (GDL), and the Universities of Stuttgart, Rome, Oxford, and Brussels.[29]

Current Advanced Supersonic Technologies

Belgium currently has no research program to develop or maintain missiles that travel at supersonic speeds. However, Belgium is interested in decreasing the cost of satellite launch and other space-related activities that require supersonic and hypersonic flight. In addition to housing the European Space Agency, the Belgian government in November 2016 stood up its own space agency with the intention of promoting its space sector.[30] Reports indicated that it intends to cooperate with China to develop a joint research program on satellite launches.[31]

Hypersonic R&D

Belgium's largest center for hypersonic R&D capability is the Von Karman Institute for Fluid Dynamics located 10 km south of Brussels. It serves as a major European testing and educational center for hypersonic technology development, though most of the research conducted by the institute itself is based in academia, with little involvement by indigenous defense companies. In 2014, the Von Karman Institute hosted a lecture series to review the comparative advances of European countries in hypersonic technology.[32]

The vast majority of Belgian involvement in hypersonic development is either as a funder of academic grants co-authored with researchers from other countries more deeply involved in hypersonic technology or as the de facto capital of the European Union. Several Belgian researchers collaborated with Stanford University and con-

---

29 Butterworth-Hayes, 2008.

30 "Belgium Plans to Create Own National Space Agency," *Sputnik News*, March 15, 2016.

31 "Belgium Plans to Create Own National Space Agency," 2016.

32 Von Karman Institute for Fluid Dynamics, "Lecture Series STO 234 on Hypersonic Flight Testing," web page, March 28, 2014.

ducted an analysis of the Australian-led HyShot II scramjet test in 2011 and 2012.[33]

Research and Testing Facilities

The Von Karman Institute in Saint Genese possesses two hypersonic wind tunnels.[34]

1. *H-3 Hypersonic Wind Tunnel*: This tunnel is capable of testing at speeds of up to Mach 6 with a test section of 12 centimeters (cm) diameter.[35]
2. *Longshot Free-Piston Gun Wind Tunnel:* Tests speeds of Mach 15 to Mach 20 for short test duration. This is a free piston tunnel with test section exit diameters of 43 cm and 60 cm.

Brazil

Current Advanced Supersonic Technology

In 2007, Brazil was named as a potential buyer of the planned Indian BrahMos I Missile—a ramjet-powered cruise missile that can reach speeds of up to Mach 3.[36]

Hypersonic R&D

Brazil began investing in hypersonic capabilities with the initiation of the 14-X hypersonic aircraft program in 2006. This experimental air-breathing jet concept would be powered first by two rockets and then propelled by an air-breathing scramjet. According to technical reports, it would have the potential to fly at Mach 7 and reach an altitude of

---

33 René Pecnik, Vincent E. Terrapon, Frank Ham, Gianluca Iaccarino, and Heinz Pitsch, "Reynolds-Averaged Navier-Stokes Simulations of the HyShot II Scramjet," *AIAA Journal*, Vol. 50, No. 8, August 2012, pp. 1717–1732.

34 Goodrich et al., 2008.

35 Von Karman Institute for Fluid Dynamics, "High Speed Wind Tunnels," web page, undated.

36 Jon Grevatt, "BrahMos Outlines $2bn Export Target for Supersonic Cruise Missile," *Jane's Defence Industry*, November 8, 2007.

30 km.[37] Development was proposed for the Brazilian Air Force's Institute for Advanced Studies.[38]

It is unclear to what extent the Brazilian government continues to fund the hypersonic project. In 2011, the first flight of the 14-X scramjet was scheduled for 2013, but papers published in 2013 indicate that the vehicle remained in the design stage.[39] In 2017, a 14-X model was tested at Mach 7–8 in preparation for a flight test not yet scheduled.[40]

Additionally, researchers at the Center for Mechanical Engineering and Applied Sciences at the Federal University of ABC in Sao Paulo, Brazil, have published academic research on hypersonic flows and air-breathing flight.[41] However, this research has been confined to feasibility studies rather than development and testing of new technologies.[42]

Research and Testing Facilities

Brazil owns and operates the T3 Hypersonic Wind Tunnel at the Air Force of Brazil General Command for Aerospace Technology in Sao Jose dos Campos.[43] Operational as of January 2007, the tunnel has a diameter of 15 cm and can test speeds of up to 8.5 km per second, or Mach 25 at sea level.[44]

---

[37] de Araújo Martos et al., 2017.

[38] Anonymous, "14-X Hypersonic Vehicle Details Given," *Flight International*, Vol. 179, No. 5287, April 2011, p. 19; Stephen Trimble, "LAAD11: Brazil Reveals Details of 14-X Hypersonic Vehicle," *Flight Global News*, April 13, 2011.

[39] Trimble, 2011. Ronaldo de Lima Cardoso and Marcos da Silva e Souza, "Brazilian 14-x S Hypersonic Scramjet Aerospace Vehicle Dimensional Design at Mach Number 7," 22nd International Congress of Mechanical Engineering (COBEM 2013), Ribeirão Preto, Brazil: November 2013.

[40] de Araújo Martos et al., 2017.

[41] de Lima Cardoso, 2013; "Propulsion and Power: New Findings on Propulsion and Power from Federal University Summarized," *Defense & Aerospace Week*, July 3, 2013, p. 396.

[42] "Propulsion and Power," 2013.

[43] Goodrich et al., 2008.

[44] Goodrich et al., 2008.

Canada

Current Advanced Supersonic Technology

While Canada did attempt to develop a supersonic plane, Avro CF-105 Arrow, rising costs and politics ultimately resulted in the termination of the project.[45] Recently, Canadian inventor and engineer Charles Bombadier revealed a design for a Mach 4 commercial plane dubbed "Skreemr."[46]

Hypersonic R&D

Canadian researchers at the University of Calgary have recently been doing some exploration of hypersonic boundary flows and errors associated with temperature and speed measurement techniques, but this work has largely been confined to computer simulations rather than physical testing.[47] To date there is little to no government investment in hypersonic technologies for space, commercial aircraft, or missile development.

Research and Testing Facilities

While Canada has a total of eight wind tunnels, they operate no hypersonic tunnels and only two supersonic tunnels.[48]

France

France sees hypersonic missile development largely as the next stage of modernization for its nuclear arsenal and essential to maintaining

---

[45] Alex Czek, "Avro Canada CF-105 Arrow Supersonic Interceptor Aircraft Prototype," *Military Factory*, August 31, 2016.

[46] "Photos: Hypersonic Jet Could Fly 10 Times the Speed of Sound," *Life Science*, November 3, 2015.

[47] "Nitrogen Oxides; Findings from University of Calgary Broaden Understanding of Nitrogen Oxides (Nitric Oxide Chemistry Effects in Hypersonic Boundary Layers)," *Defense & Aerospace Week*, 2015, p. 56.

[48] Goodrich et al., 2008.

technical parity with the United States.[49] Over the past decade, it has invested a considerable amount into R&D efforts and defense acquisitions in order to compete with the United States for the export marketplace.[50] While the commercial applications of hypersonic technology might be beneficial to the French government (in 2015, Airbus patented a design for a hypersonic jet that could travel at speeds of up to Mach 4.5 and the Mach 2+ Concorde airliner was an Anglo-French venture) and it is supportive of European efforts in space technology and design, it appears that France's primary intention in developing hypersonic capabilities is to use the technology to update its nuclear arsenal.[51]

There are two companies principally responsible for the development of French hypersonic systems and technology: the European contractor MBDA Missile Systems and France's national aerospace research center, ONERA.

Current Advanced Missile Technology

France contributes 12.4 percent of the funding toward the Meteor Missile Program (see "European Union" section in the beginning of this appendix). France additionally possesses a supersonic ramjet missile equipped to deliver nuclear payloads. The Air-Sol Moyenne Portee–A (ASMP-A) is an air-to-surface missile and travels at speeds of up to Mach 3.[52] Entered into service in 2009, ASMP-A has a range of 500 km and carries a 300-kiloton warhead.[53] The missile itself is pro-

---

49 Amy Svitak and Robert Wall, "French Legislators Push Broad Missile Defense Agenda," *Aviation Week & Space Technology,* July 18, 2011, p. 26.

50 Svitak and Wall, 2011.

51 Mary-Ann Russon, "Airbus Patents Hypersonic Jet That Can Travel from London to New York in Just One Hour," *International Business Times,* August 4, 2015; Mihai-Stefan Dinu, "Strategic Tendencies of Some Major International State Actors on 2011," *Strategic Impact,* Vol. 42, 2012, pp. 37–43.

52 Missile Threat CSIS Missile Defense Project, "Air-Sol Moyenne Portée (ASMP/ASMP-A)," web page, November 30, 2016.

53 Missile Threat CSIS Missile Defense Project, 2016.

pelled by a ramjet engine and solid fuel.[54] Originally built by Aerospatiale's Tactical Missile Division, it is now part of MBDA Missile Systems. The French have not exported the missile.

Hypersonic R&D Programs

France has been aggressively pursuing hypersonic technology since the 1990s, resulting in a number of projects that are currently in various stages of development, yet yielding promising results.

In 2000, ONERA and Aerospatiale-Matra Missiles focused their hypersonic research on strategic unmanned air vehicles (UAVs) and missile applications under a commission by the French Defense Procurement Agency (DGA).[55] Based on a Mach 8 air-launched missile, the system (dubbed "Promethee") would likely utilize a propulsion design that included a variable geometry ramjet/scramjet. The intention was to "use technology from the Franco-Russian Wide Range Ramjet program" to support the engine development.[56] In 2002, MBDA participated in a project to develop ramjet/scramjet technology for satellite launchers, but the practical applications of the project are more likely to be high-speed air-to-surface missiles and UAVs.[57]

Currently, ONERA is exploring engine technology up to Mach 8 and received a contract from the French government to develop an air-to-surface missile, known as the ASN4G, that would utilize scramjet technology to deliver nuclear payloads.[58] Defense Minister Jean-Yves Le Drian indicated in 2015 that this weapon is still decades away.[59]

Most recently, France has had success with its hypersonic technology demonstrator vehicle known as LEA—the acronym stands for

[54] Missile Threat CSIS Missile Defense Project, 2016

[55] Stanley W. Kandebo, "France, Russia to Join," *Aviation Week & Space Technology*, March 26, 2001, p. 60.

[56] Kandebo, 2001.

[57] Kandebo, 2001.

[58] "France Studies Nuclear Missile Replacement," *Defense News*, December 1, 2014, p. 22.

[59] France Studies Nuclear Missile Replacement, 2014.

the Russian phrase for "flight-test vehicle."[60] LEA conducted full-scale wind tunnel tests in 2012 under Mach 6 conditions, with flights tests that were scheduled for 2014–2015 in Russia.[61] No public information is available about the results of those tests (or whether they were in fact conducted), but the LEA vehicle is still listed by the ONERA Department of Systems Design and Performance Evaluation as an active project as of March 2017.[62]

The project is a joint venture between MBDA and ONERA, but it has also been the product of considerable collaboration with Russia. Russia's leading air-launched cruise missile developer, Raduga, has been tasked to oversee flight tests of the missile, which occur in Russia.[63] While Franco-Russo cooperative efforts were limited in the 1990s by international controls that limited the exports of engine components from Russia to France, French officials have publicly stated that work with Russia for this round will "be under contract, not on a cooperative basis."[64] In addition to MBDA and ONERA, other partners include DGA and the national research agency CNRS, Roxel, Astrium, and Auxitrol.[65] French labs are also working on problems related to hypersonic technology: The National Center for Space Studies (CNES) is actively studying regenerative cooling in an attempt to mitigate the effects of extreme heating associated with hypersonic flight.[66]

---

60 Taverna and Barrie, 2008.

61 Taverna and Barrie, 2008.

62 ONERA, The French Aerospace Lab, undated.

63 Guy Norris, "LEA Nears Delayed Hypersonic Ground Test Milestone," *Aerospace Daily & Defense Report,* September 27, 2012a.

64 Taverna and Barrie, 2008.

65 Francois Falempin and Laurent Serre, "French Flight Testing Program LEA Status in 2009," *American Institute of Aeronautics and Astronautics,* 2011.

66 J. C. Bouillot, "French Approach in Future Launch Systems," *Acta Astronautica*, Vol. 17, No. 8, August 1988, pp. 793–805.

Research and Testing Facilities

The ONERA French Aeronautics and Space Research Center contains five different hypersonic wind tunnels.[67] In 2014, ONERA requested $218 million to modernize its wind tunnel facilities at Modane and Fauga Mauzac (including hypersonic facilities).[68]

1. *R2Ch Hypersonic Wind Tunnel* (located at Chalais-Meudon): This tunnel has two separate tubes, each capable of achieving two different speeds. The first tube has a maximum range of Mach 3–4, while the second tube tests components in the range of Mach 5–7.
2. *R3Ch Hypersonic Wind Tunnel* (located at Chalais-Meudon): This tunnel can test components up to Mach 10.
3. *R5Ch Hypersonic Wind Tunnel* (located at Chalais-Meudon). This tunnel also tests up to Mach 10.
4. *F4 Arc Heated, High Enthalpy Hypersonic Wind Tunnel* (located at Fauga-Mauzac): This tunnel is composed of four tubes, each capable of producing different speed regimes. The first tube can test at Mach 8–17, the second at Mach 7–13, the third at Mach 6–11, and the final tube can achieve Mach 9–21.
5. *S4Ma Blowdown Hypersonic Tunnel* (located at Modane-Avrieux): This tunnel has three interchangeable nozzles. The first achieves speeds at Mach 6.4, the second Mach 10, and the third Mach 21.

Germany

Germany currently appears to be focusing on pan-European projects. Because of Germany's position as the economic leader in the European Union, German decisionmaking and interests heavily influence

---

[67] Goodrich et al., 2008; ONERA, The French Aerospace Lab, "Hypersonic Wind Tunnel," web page, 2009.

[68] Pierre Tran, "Onera Explores Mach-8 Missile Engine Technology," *Defense News*, October 19, 2015, p. 17.

many of the defense positions and technological advances in the European Union. The Supersonic and Hypersonic Transport Department at German aerospace company DLR is a principal contributor to multiple European Union-funded research activities, while the German contractor Bayern-Chemie is a leader in the production of ramjets and other high-speed engines.

Current Advanced Supersonic Technology

Germany provides 16 percent of the funding of the Meteor Missile Program (see "European Union" earlier in this appendix).

Hypersonic R&D

Germany first committed significant funding to research in hypersonic technology in 2000 through a joint partnership with Sweden (a project named HFK), but it pulled funding in 2003 despite a series of successful tests.[69]

Shortly after pulling funding for the joint project with Sweden, Germany aerospace company DLR invested considerable time and materials into the development of the Sharp Edge Flight Experiment (SHEFEX) research program, which ultimately produced a hypersonic glide vehicle demonstrator first tested in 2005 (SHEFEX I) and again in the summer of 2012 (SHEFEX II).[70] After a successful test in summer 2012 reached Mach 11, plans were made to test a small proof-of-concept suborbital RV, but funding appears to have stopped for the program sometime after DLR published a review of temperature measurements from the SHEFEX II test in December 2014.[71] A previously planned 2016 test has yet to be announced and the project has disappeared from the DLR publicity regarding ongoing research.

---

69 "Bruisers, Incoming: Anti-ship Strike Takes and Eastwards Fix," *IHS Jane's Navy International,* October 16, 2013.

70 Michael Dumiak, "North Star," *Aviation Week & Space Technology,* Vol. 174, No. 32, September 3, 2012.

71 "Spacecraft and Rockets; Data on Spacecraft and Rockets Reported by Researchers at German Aerospace Center (Transpiration-Cooled Hypersonic Flight Experiment: Setup, Flight Measurement, and Reconstruction)," *Defense & Aerospace Week,* 2015, p. 91.

Currently, German defense contractor DLR is an active partner in both the LAPCAT II and ATLLAS II programs (see "European Union" earlier in this appendix).

Research and Testing Facilities

Germany hosts at least three hypersonic wind tunnel facilities.[72]

1. *The Hypersonic Ludwieg Wind Tunnel (HLB)*: Housed at the Carolo-Wilhelmina Technical University at the Braunschweig Institute for Fluid Mechanics, it has a 0.5 m diameter test section and is capable of testing speeds of up to Mach 6.[73]
2. *The Intermittent Ludwieg Tube Wind Tunnel (RWG)*: Based at the German-Dutch Wind Tunnels in Gottingen, Germany, it contains two legs of 0.5 m diameters each. The RWG is used primarily for space vehicle and missile R&D, capable of achieving speeds of Mach 2.9–4.65 (first tube) and Mach 5.0–6.9 (second tube).[74]
3. *H2K Hypersonic Wind Tunnel at the Institute of Aerodynamics and Flow Technology*: Operated by DLR at their Cologne facility, this is a "blow-down" wind tunnel capable of testing speeds of up to Mach 11.2.[75]

India

India's Defense Research and Development Organization (DRDO) currently has two parallel programs in hypersonic development, each making considerable strides toward being operational. Their collaborative research program, BrahMos II, is a joint venture with Russia

---

[72] Goodrich et al., 2008; Institute of Aerodynamics and Flow Technology, "Hypersonic Wind Tunnel Cologne (H2K)," web page, undated.

[73] Goodrich et al., 2008.

[74] Goodrich et al., 2008.

[75] Institute of Aerodynamics and Flow Technology, undated.

that expects to produce a hypersonic cruise missile capable of reaching Mach 7 by 2017. The indigenous research project, the HSTDV, planned to conduct its first flight test in late 2016 or early 2017. Both the BrahMos II and the HSTDV will be used as missiles to carry warheads. Additionally, the Indian Space Research Organization (ISRO) is sponsoring R&D on an air-breathing reusable launch vehicle–technology demonstrator as a first step toward developing a two-stage-to-orbit reusable launch vehicle.

Current Advanced Supersonic Technology

BrahMos I

India's cooperative venture with Russia led to the development and production of the BrahMos I Missile, a cruise missile capable of reaching speeds of up to Mach 3.[76] While India is technically the senior stakeholder, Russia owns 49.5 percent of the company, and the seeker technology used for guiding the missile is exclusively Russian, making the project dependent upon Russian cooperation and willingness to share technology.[77] After a temporary hold on exports, India has agreed to export the BrahMos I to such countries as Vietnam, South Africa, Brazil, Chile, and the United Arab Emirates. India was additionally involved in discussions over potential BrahMos I contracts with Philippines, South Korea, Algeria, Greece, Malaysia, Thailand, Egypt, Singapore, Venezuela, and Bulgaria.[78]

Reusable Launch Vehicle Technology Demonstrator

The ISRO has been investing in an air-breathing technology demonstrator that would travel at speeds of up to Mach 5 and set the stage for

---

76 "Military and Technical Co-Operation: BrahMos Missile Air-Based Modification to Be Ready by Late 2014," *Interfax: Russia & CIS Military Information Weekly*, March 1, 2013; Rahul Bedi, "DRDO to Build New Missile Test Range on Southeast Coast of India," *Jane's Defence Weekly*, Vol. 49, No. 51, November 28, 2012.

77 Arming India, "India: BrahMos Aerospace Chief Comments on Ambition to Achieve 90-Degree Steep Dive Capability," interview with Sudhir Kumar Mishra, March 27, 2016.

78 Shiv Aroor, "BIG! Indian Govt Clears Decks for BrahMos Export," *Livefist Defence*, June 3, 2016; Grevatt, 2007.

a more cost-efficient method of space launch. The two-stage-to-orbit reusable launch vehicle has an estimated ability to decrease the cost of a space launch by a factor of ten, and as a result the ISRO is very interested in its development.[79] A first-design technology demonstration test was conducted on May 23, 2016—a ten-minute flight of a subscale vehicle that was deemed successful by the agency.[80]

Hypersonic R&D

BrahMos II

The BrahMos II is the next generation of cruise missile being designed by Indian and Russian engineers and financed by both governments. The missile is expected to have a range of 290 km and be capable of carrying a warhead up to 300 kg in weight—a range that puts Islamabad within reach from Indian territory.[81] Russia currently contributes $122 million to the program, while India is contributing $128 million to the R&D effort.[82] It is unclear whether the 2017 completion goal is realistic, given the lack of completed test flights to date. Both Russian and India have stated that they do not plan to export the BrahMos II missile.[83]

Hypersonic Demonstrator Vehicle

The HSTDV is India's indigenous effort at building a hypersonic cruise missile using scramjet technology. It is separately researched and funded from BrahMos II and entirely indigenous. The program's goal is to develop a scramjet capable of speeds of up to Mach 7 and a hypersonic cruise missile altitude of 20 miles.[84] Although the program's first

---

[79] "Indian Reusable Launch Vehicle Testbed Ready for High-Speed Flight," *Spaceflight 101*, May 21, 2016.

[80] Kiran Krishnan Nair, "Apples and Oranges: Why Comparing India's Reusable Launch Vehicle with the Space Shuttle Is Totally Out of Place," *The Space Review*, May 23, 2016.

[81] Uebler, 2012, p. 128.

[82] Uebler, 2012.

[83] Uebler, 2012.

[84] Menon, 2012, p. 51.

test flight was planned for late 2016–2017, the program has consistently failed to meet scheduled milestone goals.[85] It scheduled a critical 20-second burn test for the end of 2016.[86]

The program's first wind tunnel test was conducted in 2007 in Israel and the second in Russia in 2009, as India lacked a testing facility with a sufficient cross section.[87] India recently completed construction of a hypersonic wind tunnel, inaugurated in April 2014, which will fill some gaps in their testing capabilities.[88]

Other Collaborative Efforts

India established a scientific and technical cooperative effort with Belarus, with agreements to share technologies and research in a variety of fields, including the "development of Background-Oriented Schlieren (BOS) technique for hypersonic flow field diagnostics," suggesting that the two countries may be cooperating in understanding the aerodynamic flows around hypersonic vehicles.[89]

Research and Testing Facilities

India's newest wind tunnel was inaugurated on April 8, 2014, at the Indian Institute of Science and was designed to conduct tests for the HSTDV project.[90] It is estimated that the defense industry will account for 60 percent of the wind tunnel's use. The ISRO runs another facility on the property.[91] In total, India has at least 12 hypersonic wind tunnels across the country.

---

85 Robert Hewson, "India's DRDO Preparing for Hypersonic Test," *Jane's Defence Weekly*, Vol. 49, No. 46, October 24, 2012b.

86 "Indian Reusable Launch Vehicle," 2016.

87 Archit Gupta, "Hypersonic Aircraft at Mach 6.5 powered by DRDO2014," *Indian Aviation News*, October 23, 2014.

88 Menon, 2012; Kelvin Wong, "India Opens New Hypersonic Wind Tunnel Facility," *Jane's International Defense Review*, Vol. 47, No. 5, May 1, 2014.

89 Embassy of the Republic of Belarus in the Republic of India, "Scientific and Technical Cooperation," web page, undated.

90 Wong, 2014.

91 Wong, 2014.

Indian Institute of Science—Bangalore

1. *0.3 m Hypersonic Wind Tunnel:* Intermittent blowdown tunnel that tests speeds of Mach 5.4–10.2. Test section diameter of 0.3 m.[92]
2. *0.5 m Hypersonic Wind Tunnel*: Tests speeds of up to Mach 8.0, test section diameter of 0.5 m.[93]
3. *300 x 300 mm Hypersonic Shock Tunnel HST2*: Tests speeds of Mach 6–12. It is used to generate force, pressure, and heat transfer data for high-enthalpy studies and temperatures up to 5,000 K.[94]
4. *Free Piston Driven Hypersonic Shock Tunnel HST3*: Tests speeds of Mach 6–12 using 0.3 m diameter test section.[95]
5. *Hypersonic Shock Tunnel HST4*: Tests speeds of Mach 6–12 using 2.5 m-long cylindrical tank of 1.5 m diameter. It can withstand specific enthalpy of 3 megajoule (MJ)/kg.[96]
6. *Hypersonic Shock Tunnel HST5*: Established in 2008, can test speeds of Mach 6–12 with specific enthalpy of 6 MJ/kg. [97]

---

[92] Aerospace Testing Facilities Available in India, "Title of the Aerospace Test: 0.3m Hypersonic Wind Tunnel," web page, undated-b.

[93] Aerospace Testing Facilities Available in India, "Title of the Aerospace Test: 0.5m Hypersonic Wind Tunnel," web page, undated-c.

[94] Goodrich et al., 2008; Aerospace Testing Facilities Available in India, "Title of the Aerospace Test: 0.5m Hypersonic Shock Tunnels HST2," web page, undated-d.

[95] Aerospace Testing Facilities Available in India, "Title of the Aerospace Test: Free Piston Driven Hypersonic Shock Tunnel HST3," web page, undated-g.

[96] Aerospace Testing Facilities Available in India, "Title of the Aerospace Test: Hypersonic Shock Tunnel HST4," web page, undated-h.

[97] Aerospace Testing Facilities Available in India, "Title of the Aerospace Test: Hypersonic Shock Tunnel HST5," web page, undated-i.

Vikram Sarabhai Space Center—Thiruvandrum

7. *0.25 m Hypersonic Wind Tunnel* (ISRO): Established in 1980 and capable of testing speeds of Mach 4–8 at temperatures up to 700 K; test section diameter of 0.25m.[98]
8. *Hypersonic Wind Tunnel* (ISRO): Completed in 2012 with a test section size of 1 m diameter and capable of testing speeds of Mach 6–12, it is responsible for most of the ISRO's reusable launch vehicle testing and development.[99]
9. *0.3 m Hypersonic Shock Tunnel* (ISRO): Established in 1980, it is capable of testing speeds of Mach 6–10 at temperatures up to 3,000 K with a test section size of 0.3 m diameter.[100]
10. *Hypersonic Shock Tunnel* (ISRO): Capable of testing speeds of up to Mach 13 (4.5 km per second) at temperatures up to 5,000 K.[101] It became operational in 2012.

Defence Research and Development Laboratory (DRDL)—Hyderabad

11. *DRDL Hypersonic Shock Tunnel:* High-speed intermittent wind tunnel; tests speeds of Mach 6, 6.5, 7, 8, 9, and 10 up to temperatures of 4,000 K.[102]

Indian Institute of Technology—Chennai

12. *Combustion Driven Shock Tunnel*: Built in 1986 and capable of testing speeds of Mach 5–12 at temperatures up to 800 K, it is a

---

[98] Aerospace Testing Facilities Available in India, "Title of the Aerospace Test: 0.25m Hypersonic Wind Tunnel," web page, undated-a.

[99] Wong, 2014; Aerospace Testing Facilities Available in India, "Title of the Aerospace Test: 1 m Hypersonic Wind Tunnel," web page, undated-e.

[100] Aerospace Testing Facilities available in India, undated-b.

[101] Aerospace Testing Facilities Available in India, "Title of the Aerospace Test: Hypersonic Shock Tunnel," web page, undated-f.

[102] Aerospace Testing Facilities available in India, undated-f.

pressure-vacuum-type tunnel and responsible for both institute research and national programs.[103]

Iran

Iran has been investing in both its nuclear program and its missile technologies for the last two decades, and modernizing its missile arsenal appears to be a top priority for the Revolutionary Guard—Iran's top military force with a powerful political presence. Iran completed construction on its first hypersonic wind tunnel in March 2014—a step forward in its ability to develop, test, and protect its indigenous missile R&D, including work on hypersonic vehicles.

Current Advanced Supersonic Technology

Iran has stated its intention to build anti-ship missiles capable of traveling at supersonic speeds.[104] In August 2016, while showcasing the new Bavar-373 air defense missile, the Iranian Minister of Defense Brigadier General Hussein Dehqan revealed a turbojet that he claimed "would be used to develop a supersonic cruise missile in the near future."[105] To date, however, Iran does not appear to be developing these capabilities.[106]

Hypersonic R&D

Academic research from the past four years complements Iran's interest in hypersonics, with scholars from the University of Tehran pursuing research on hypersonic transitional flows, researchers from the

---

[103] Goodrich et al., 2008; Department of Aerospace Engineering: Indian Institute of Technology Madras, "Research—Aerodynamics," web page, undated.

[104] "Iran to Build Supersonic Maritime Cruise Missiles Soon: Dehqan," *PRESSTV News*, August 21, 2016.

[105] Leith Fadel, "Iran to Develop Supersonic Cruise Missile," *AMN Al Masdar News*, August 22, 2016.

[106] Farzin Nadimi, "Iran and China Are Strengthening Their Military Ties," *The Washington Institute*, November 22, 2016.

Amirkabir University of Technology in Tehran evaluating flows at Mach 6, and faculty at the Babol University of Technology identifying how vehicles traveling at hypersonic speeds could benefit from counter-flowing jets as a cooling system upon reentry.[107]

Research and Testing Facilities

Iran's newest facility, located at the University of Tehran, is a hypersonic wind tunnel capable of testing speeds of up to Mach 8—almost three times faster than its previous tunnel.[108] According to news reports, it will provide Iran significant advantages in testing and development of hypersonic technologies, including the ability to keep costs low and keep technology proprietary and inside the country. With the new facility, Iran no longer has to send its designs outside of the country for testing, saving costs and protecting proprietary technology that may provide insights into the status of the research.[109]

Israel

Current Advanced Supersonic Technology

There were some unconfirmed reports that a new supersonic anti-ship missile tested in March 2016 was ramjet powered.[110] The surface-to-

[107] "Aerospace Research: Reports from University of Tehran Advance Knowledge in Aerospace Research," *Defense & Aerospace Week*, November 7 2012, p. 389; "Aeronautics and Astronautics: New Aeronautics and Astronautics Study Results Reported from Amirkabir University of Technology," *Defense & Aerospace Week*, July 11, 2014a, p. 76; "Aerospace Research: Recent Research from Babol University of Technology Highlight Findings in Aerospace Research (Heat Reduction Using Counterflowing Jet for a Nose Cone with Aerodisk in Hypersonic Flow)," *Defense & Aerospace Week*, February 25, 2015, p. 130.

[108] Umid Niayesh, "Iran Builds First Hypersonic Wind Tunnel to Test Missiles and Spacecraft," *Trend News Agency*, March 5, 2014.

[109] Niayesh, 2014.

[110] "Israel's Navy New Anti-Ship Missile," *Defense Update News*, May 27, 2016.

surface missile, designated Gabriel V, is reportedly being developed by Israel Aerospace Industries (IAI) to replace legacy Gabriel II missiles.[111]

Hypersonic R&D

There is little evidence that Israel is actively engaged in the development of a hypersonic research program for space, commercial, or defense purposes. Some reports suggest that Israel Military Industries (IMI) is developing propulsion systems for scramjet-powered hypersonic vehicles, but it is unclear how many resources—academic or government-sponsored—are being committed to indigenous hypersonic technologies.[112]

Research and Testing Facilities

Israel has two hypersonic wind tunnel facilities that have been used by other countries such as India and Japan for testing of spaceplane designs and materials.[113]

1. *Arc Plasma Generator/Hypersonic Wind Tunnel* (located at the Technion-Israel Institute of Technology, Haifa): It can test designs up to Mach 8 at temperatures up to 6,000 K.[114]
2. *IAI Hypersonic Wind Tunnel* (located in Lod, Israel): Established in 1989, it can test speeds at Mach 5, 6, 8, 10, and 12 with a test section size of 0.45 m in diameter.[115]

---

[111] "Strategic Weapon Systems," *Jane's Sentinel Security Assessment–Eastern Mediterranean*, February 7, 2017.

[112] "Israel Military Industries Ltd (IMI)," *Jane's Space Systems and Industry*, March 7, 2007.

[113] Arie Egozi, "Israel Helps India Develop Scramjet Demonstrator," *Flight International*, July 11, 2007.

[114] Goodrich et al., 2008.

[115] IAI: Israel Aerospace Industries, "Wind Tunnels Center," web page, undated.

Italy

Italy's intentions regarding hypersonic technical development appear to be almost exclusively commercial, with very little (if any) government demand for defense applications. While Italy has partnered with France to improve European ballistic missile defense through the Surface to Air Missile Platform/Terrain (SAMP/T) program, it lacks the kind of geostrategic use for hypersonic weapons that France or the UK envisions.

Current Advanced Supersonic Technology

Italy contributes 12 percent of the funding for the Meteor Missile Program (see "European Union" earlier in this appendix). As of early 2017, the missile is not yet in service with the Italian Air Force.[116]

Hypersonic R&D

Italy has a host of civil organizations and academic institutions conducting research on commercial hypersonic technologies, but little formal government sponsorship of defense-related research to develop hypersonic weapons. The principal agency responsible for developing hypersonic technologies in Italy is the Italian Aerospace Research Center (CIRA), a private consortium company created to manage the Italian Aerospace Research Program (PRORA), of which the government owns 68 percent of the shares.[117] It is currently a major partner in the LAPCAT II program, evaluating engines that could travel at speeds of Mach 5 and Mach 8.[118] Finally, CIRA also administers the Italian National Propulsion program HYPROB, which has a goal of evolving and consolidating national technology and system development capabilities on rocket propulsion for future space applications.[119]

---

[116] Richard Tomkins, "Swedish AF Gripens Now Carry Meteor Missiles," *Defense News*, July 11, 2016.

[117] Italian Aerospace Research Centre, "Cira and Prora," web page, undated-a.

[118] Italian Aerospace Research Centre, "LAPCAT II: Long-Term Advanced Propulsion Concepts and Technologies," web page, undated-d.

[119] Italian Aerospace Research Centre, "HYPROB," web page, undated-c.

Academic research on hypersonic technology in Italy is widespread, with researchers in industry, the University of Naples, Polytechnic of Turin, University of Rome, and other institutions performing preliminary performance studies of small prototypes, plasma effects around RVs, and pre-feasibility studies on high-altitude flight.[120]

Research and Testing Facilities

CIRA runs two hypersonic wind tunnels used for testing spacecraft reentry conditions.

1. *Scirocco Plasma Hypersonic Wind Tunnel* (located in Capua, Italy): This is a hypersonic, thermo-structural tunnel with an electric arc heater that has a maximum power of 70 megawatts. It is capable of testing objects at speeds of up to Mach 12. It is typically used to focus on the development and qualification of thermal protection systems for aerospace use.[121]
2. *GHIBLI Hypersonic Plasma Tunnel* (located in Capua, Italy): This is a smaller tunnel, but can test models up to 8 cm in diameter.[122]

---

[120] "Aeronautics and Astronautics: New Findings from Marche Polytechnic University in the Area of Aeronautics and Astronautics Described," *Defense & Aerospace Week*, June 12, 2013, p. 320; "Aeronautics and Astronautics: Findings from F. Santoro et al Update Understanding of Aeronautics and Astronautics (the Italian Spacegate: Study and Innovative Approaches to Future Generation Transportation Based on High Altitude Flight)," *Defense & Aerospace Week*, July 30, 2014b, p. 159; "Alkenes: Report Summarizes Alkenes Study Findings from Q.L. Liu and Co-Researchers (Ethylene Flame Dynamics and Inlet Unstart in a Model Scramjet)," *Defense & Aerospace Week*, December 10, 2014, p. 122; "Aeronautics and Astronautics: Researchers from Technical University Report on Findings in Aeronautics and Astronautics Performances of a Small Hypersonic Airplane (HyPlane)," *Defense & Aerospace Week*, September 16, 2015, p. 145; "Spacecraft and Rockets; Researchers from University of Naples Federico II Describe Findings in Spacecraft and Rockets (Plasma Effect on Radiofrequency Communications for Lifting Reentry Vehicles)," *Defense & Aerospace Week*, April 29, 2015, p. 146.

[121] Goodrich et al., 2008; Italian Aerospace Research Centre, "Facilities," web page, undated-b.

[122] Italian Aerospace Research Centre, undated-b.

Japan

Current Advanced Supersonic Technology

Japan's Technical Research and Development Institute is currently developing a supersonic anti-ship missile, called the XASM-3, in collaboration with Mitsubishi Heavy Industries. The missile will integrate a solid-fuel rocket with a ramjet capable of operating at speeds of up to Mach 5. It is expected to have a range of more than 120 miles.[123]

Additionally, the Zehst program sponsored by Airbus has ties with Japan. These plans are for a 60 passenger near-hypersonic vehicle that would be turbo-jet powered and reach speeds between Mach 4 and Mach 5, with a commercial vehicle planned for 2050.[124] It is unclear how developed these plans really are, or what kind of research and funding they have generated. Some officials have stated a desire to integrate the Zehst program with other supersonic/hypersonic initiatives like the Hikari program (described in the following section).[125]

Hypersonic R&D

The primary organization responsible for hypersonic research in Japan is JAXA. Formed in 2003 with the merger of the National Space Development Agency, the Institute of Space and Aeronautical Science, and the National Aeronautics Laboratory, it is involved in a series of research programs related to the modeling and testing of hypersonic technologies, with the ultimate goal to create a hypersonic airliner.[126] In 2005, it released the JAXA Vision/JAXA 2025 project, which details JAXA's

---

[123] Kyle Mizokami, "Bullseye: The 5 Most Deadly Anti-Ship Missiles of All Time," *The National Interest*, March 13, 2015; Navy Recognition, "Naval Forces News–Japan," web page, November 16, 2015.

[124] Guy Norris, "Europe and Japan Forge New Hypersonic Links," *Aviation Week & Space Technology*, November 26, 2012b; "Europe, Japan Closer to Hypersonics Tech Plan," *Aerospace America*, May 2014, p. 5; Guy Norris, "Polar Express," *Aerospace America*, web page, May 2014.

[125] Norris, 2014.

[126] Frank Morring Jr. and Michael Mecham, "'One JAXA'," *Aviation Week & Space Technology*, Vol. 163, No. 21, November 28, 2005, p. 64.

mission to develop a hypersonic aircraft that can cruise at Mach 5 and cross the Pacific Ocean in two hours.[127]

As a part of this vision, Japan is invested in hypersonic research as a partner in the Hikari program (High Speed Technologies for Future Air Transport Research and Innovation). Together with the European Commission, the Japanese Ministry of Economy, Trade, and Industry co-founded the Hikari program to search for a joint approach to perfecting key propulsion-related technologies.[128] The Hikari program directors hope to begin experimentation for a future hypersonic vehicle by 2020. The project to date has completed an evaluation of a common technology baseline for the vehicle with cost-benefit calculations to determine economic feasibility for a hypersonic X-plane by 2025.[129]

To date, JAXA has developed a conceptual model for a hypersonic aircraft capable of cruising at Mach 4.5 and traveling trans-Pacific routes in just a few hours.[130] This project, called Hypersonic Technology Experimental Aircraft (Hytex), would use dual precooled liquid hydrogen-fueled turbojets, but it is still in early development.[131] A flight test, yet to be approved, may use a two-stage solid fuel rocket to release the vehicle at Mach 5.[132]

Universities in Japan are also investigating hypersonic technologies. The University of Tokyo's Department of Advanced Energy published a paper on the aerothermodynamics of hypersonic airflows, while researchers at Nihon University simulated the acoustic properties of high-temperature, high-velocity jets from a rectangular hypersonic nozzle.[133]

---

[127] "JAXA 2025 (JAXA Long-Term Vision)," 2009.

[128] Loctier, 2015; Norris, 2012b; Norris, 2014f.

[129] Norris, 2012b.

[130] Guy Norris, "Turbojet Test," *Aviation Week & Space Technology*, November 26, 2012c.

[131] Brian Wang, "Japan and Europe Cooperating to Develop the Technology for Hypersonic Commercial Passenger Planes," *Next Big Future*, March 9, 2015; Taguchi, Marakami, Sato, and Tsuchiya, 2009, pp. 27–32.

[132] Norris, 2012c.

[133] "Propulsion and Power," 2013; "Propulsion and Power; New Findings from Nihon University in the Area of Propulsion and Power Described (Acoustic Simulation of Hot Jets Issu-

Research and Testing Facilities

Japan has multiple hypersonic wind tunnels at various locations, including Japan Aerospace Exploration Agency, Mitsubishi Heavy Industries, and the University of Tokyo.

Japan Aerospace Exploration Agency

1. *0.5 m Hypersonic Wind Tunnel (HWT1)*: Capable of testing speeds of Mach 5, 7, and 9 with interchangeable nozzles and a test section of 0.5 m in diameter. It was completed in 1965.[134]
2. *1.27 m Hypersonic Wind Tunnel (HWT2):* Tests speeds of up to Mach 10 with a test section size of 1.27 m in diameter and a fixed nozzle. It was completed in 1995.[135]
3. *0.44 m Hypersonic Shock Tunnel*: Capable of testing speeds of up to Mach 12 with a test section size of 0.44 m in diameter.[136]

Mitsubishi Heavy Industries

4. *High Enthalpy Shock Tunnel (HIEST)*: Free-piston tunnel used for testing scramjet engines and spacecraft, with a shock tube diameter of 0.18 m.[137] It tests speeds of up to Mach 10 and Mach 12.[138]

University of Tokyo (Kashiwa Campus)

5. *20 cm Hypersonic Wind Tunnel*: Tests speeds at Mach 7 and Mach 8 up to temperatures of 1,000 K. Test section size of

---

ing from a Rectangular Hypersonic Nozzle)," *Defense & Aerospace Week*, July 2, 2014, p. 121.

[134] Wind Tunnel Technology Center, "Hypersonic Wind Tunnels," web page, undated.

[135] Wind Tunnel Technology Center, undated.

[136] Goodrich et al., 2008.

[137] Mitsubishi Heavy Industries, "High Enthalpy Shock Tunnel (HIEST)," web page, undated-b.

[138] Goodrich et al., 2008.

0.2 m in diameter, constructed by Mitsubishi Heavy Industries in 1965.[139]

The Netherlands

The Netherlands government has expressed little interest in funding or developing hypersonic capabilities, other than its collaboration with German research interests. Its intent in funding wind tunnel testing appears to be primarily academic in nature. While the state is heavily invested in aerospace—both through its partnerships with German aeronautics and because the Airbus Group (formerly known as EADS) is headquartered in Leiden, Netherlands—its resources appear principally dedicated to subsonic and transonic research.

Current Advanced Supersonic Technology

In 2007, the Netherlands Organization for Applied Scientific Research collaborated with the Swiss-based contractor RWM Schweiz to develop and test a ramjet-powered projectile that could be fired from a gun as ammunition. The solid fuel ramjet (SFRJ) projectile would increase the range, speed, and kinetic energy of current ammunition, traveling at speeds as high as Mach 4.[140]

Hypersonic R&D

The Netherlands currently has few indigenous hypersonic capabilities. Its contributions to hypersonic development are either academic or space-related through the European Space Research and Technology Center, headquartered in Noordwijk, Netherlands. This "incubator of the European Space effort" contains a testing center for satellites and is currently focused on advancing capabilities for vehicle reusability and

---

139 Mitsubishi Heavy Industries, "20 cm Hypersonic Wind Tunnel," web page, undated-a; University of Tokyo, "Hypersonic and High Enthalpy Wind Tunnel Kashiwa Campus, University of Tokyo," Graduate School of Frontier Sciences (GSFS) Division of Transdisciplinary Sciences, June 2006.

140 "TNO, RWM Create New Class of Projectile," *Jane's Missiles and Rocket,* February 1, 2007.

reentry.[141] In February 2015, it successfully launched IXV (see "European Union" earlier in this appendix).[142]

Research and Testing Facilities

The Delft University of Technology has 11 high- and low-speed wind tunnels for aerodynamic testing, including a hypersonic tunnel that can accommodate speeds of up to Mach 11.[143] Based on a Ludwieg Tube principle, it typically investigates double compression ramps, hypersonic capsule aerodynamics, the development of Particle Image Velocimetry for hypersonic flows and roughness-induced boundary layer transition.

The University of Delft boasts a hypersonic wind tunnel for its High Speed Laboratory, where researchers collaborate with a variety of institutions to evaluate hypersonic aerodynamic flows. At present, it is sponsoring projects that collaborate with Fokker Aerostructures and Fokker Elmo (a Dutch aerospace company that principally develops landing gear and electrical services for the aerospace and defense industry), the German Aerospace Laboratory (DLR, which is actively involved in research around hypersonic missiles), the Israel Institute of Technology (Technion), and a growing relationship in education and research with the Chinese aviation industry.[144]

---

141 ESTEC: European Space Research and Technology Centre "About Us," web page, February 19, 2016.

142 The vehicle is designed to reach low-Earth orbit heights, but never in fact makes a full rotation around the Earth. It is intended to be a reusable satellite launch vehicle that is able to reenter the Earth's atmosphere after reaching a maximum altitude of 256 miles. See Tate, 2015.

143 Goodrich et al., 2008; "HTFD Hypersonic Wind Tunnel (HSL)," *Delft University of Technology*, undated.

144 Fokker, 2006.

Norway

Current Advanced Supersonic Technology

In 2015, it was reported that the U.S. Naval Air Warfare Center Weapons Division's (NAWCWD's) T-Range thermal dynamics test facility was currently supporting an expected partnership between the U.S. and Norway to develop and test solid-fuel ramjets to begin in 2017.[145] It is unclear whether that partnership is currently under way.

Hypersonic R&D

Norway's principal contribution to hypersonic missile development is its rocket range and testing facility in Andoya. There is no indigenous hypersonics R&D program in Norway.

Research and Testing Facilities

Norway maintains one of the premier test centers in the world at the Andoya Rocket Range and Test Center. Ninety percent owned by the Norwegian Department of Trade and Industry, and 10 percent held by Kongsberg Defense Systems, the Andoya Space Center is a commercial test range that is open to international clients.[146] It launches test flights for hypersonic missiles and prototypes, and is primarily competitive with the Woomera test site in South Australia. In 2015, the joint U.S.-Australian project HIFiRE conducted tests at Andoya because the Woomera Range was unavailable.[147] Clients of the test range include the European Space Agency, NASA, JAXA, and multiple international universities and institutes seeking to test rocket design, system updates, and new propulsion systems.[148]

---

145 Doug Richardson, "NAWCWD Upgrades Its Sea-Level Engine Test Facility," *Jane's Missiles & Rockets*, May 1, 2015.

146 Andoya Space Center (ASC), "About Us," web page, undated-a.

147 Australian Government Department of Defence, "Valuable Lessons Learnt in Latest Hypersonic Flight Trials," August 12, 2015.

148 Andoya Space Center, undated-a; Andoya Space Center (ASC), "Current GCI Missions," web page, undated-b.

Pakistan

Current Advanced Supersonic Technology

In 2013, China announced that Pakistan was to be the first export customer of the new CM-400AKG supersonic ramjet-powered anti-ship missile.[149] Imports of this radar-guided missile allow Pakistan to strike targets up to 500 km away at speeds of up to an estimated Mach 4. Reports indicate that the missile is currently in service with the Pakistani Air Force.[150]

Hypersonic R&D

Pakistan currently does not appear to have a program to develop a hypersonic cruise missile or research scramjet technology. While Pakistan has some indigenous capabilities regarding missile development, it also relies on Chinese exports for much of its short- and medium-range ballistic missile demands.[151]

Research and Testing Facilities

Pakistan operates one subsonic wind tunnel test facility at the National University of Science and Technology in Risalpur and two other wind tunnels of unknown capabilities that test missile and projectile development.[152] Our literature review did not reveal any hypersonic wind tunnel test facilities.

---

149 "YJ-12 (CM-302)," 2016.

150 Stephen Trimble, "DUBAI: China Details Performance of 'Carrier Killer' Missile for JF-17," *Flight Daily News*, November 19, 2013.

151 "YJ-12 (CM-302)," 2016.

152 Goodrich et al., 2008.

Singapore

Current Advanced Supersonic Technology

Singapore does not appear to be developing an indigenous capability for ramjet propulsion systems or supersonic missiles. Nor does it appear to be interested in importing such weapons.

Hypersonic R&D

Singapore's strategic position in global trade and regional aviation resulted in it being selected for a prize in innovative designs for hypersonic travel that was awarded in December 2008.[153] The country has an office of engineering research consultants incorporated in 2002 that focuses on hypersonic engineering, but there is little evidence of activity.[154] Beyond some interest in Singapore as a hub for hypersonic airline activities, however, it has no hypersonic R&D program to speak of.

Former Ministry of Defense employee Lim Seng, the EADS (now Airbus) chief technology officer for Asia Pacific, set up the Singapore EADS R&D lab in 2010 and initiated the Space Plane Demonstrator project in Singapore. This project, sponsored by Airbus and run by Singapore firm Hope Technik, is currently determining the feasibility of a four-person commercial space plane.[155]

Research and Testing Facilities

Singapore operates one supersonic wind tunnel facility at the National University of Singapore. It is primarily used by the Ministry of Defense for R&D in aerodynamics and by faculty/staff members of the mechanical engineering department.[156]

---

[153] T. Scott Morley, "From Singapore to Sydney: A Prize for Hypersonic Point to Point Transportation," International Space University, March 25, 2008.

[154] InfoLabel, "Engineering—Research Consultants (Miscellaneous) in Singapore," web page, undated.

[155] ITU: The International Telecommunication Union, "Biography of Lim Seng," web page, undated.

[156] Goodrich et al., 2008.

South Korea

Current Advanced Supersonic Technology

South Korea is reported to be developing a ramjet-powered missile, the Haeseong-2 land-attack cruise missile. Unconfirmed reports suggest that the missile was tested a dozen times between 2007 and 2009, has a range of 500 km, and was scheduled for deployment in 2013.[157] However, we have seen no new information about the missile since the *Korean Times* broke the story in 2011.[158]

Hypersonic R&D

The principal research on hypersonic flows in South Korea being reported today is being conducted at the university level in academic settings. Researchers at Pusan National University, Seoul National University, and the Korea Advanced Institute of Science and Technology have all published research that models hypersonic flows and investigates the aerothermodynamics of hypersonic flight.[159] However, this research has largely been confined to computational modeling rather than testing, as South Korea lacks the appropriate wind tunnel facilities to develop hypersonic flight vehicles. Some reporting has asserted that South Korea has a dedicated hypersonic program, but the evidence suggests that the South Korean government's priorities lie with missile defense and supersonic technology rather than hypersonic capabilities.[160]

---

157 "Haeseong (SSM-700K)," *Jane's by IHS Markit: Weapons: Naval*, March 4, 2017.

158 Kalyan M. Kemburi, "High-Speed Cruise Missiles in Asia: Evolution or Revolution?" *Fair Observer*, March 19, 2014.

159 "Aerospace Research; Study Results from Pusan National University in the Area of Aerospace Research Reported (Multidimensional Flux Difference Splitting Schemes)," *Defense & Aerospace Week*, September 23, 2015, p. 134; "Aerospace Research: Findings from Seoul National University Yields New Data on Aerospace Research," *Defense & Aerospace Week*, December 11, 2013, p. 61; Gisu Park, "Study of Oxygen Catalytic Recombination," 43rd AIAA Thermophysics Conference, Fluid Dynamics and Co-located Conferences, New Orleans, La., June 25, 2012.

160 Nayef Al-Rodhan, "Hypersonic Missiles and Global Security," *The Diplomat*, November 13, 2012.

South Korea is actively building up its space program and completed a successful orbit in January 2013. South Korea is currently developing an indigenous rocket with plans to launch its first lunar orbiter in 2023.[161] Its current ship-launched Haeseong missile arsenal is capable of hitting any point inside of North Korea.

Research and Testing Facilities

South Korea is not reported to have any major hypersonic wind tunnel testing facilities.

Spain

Current Advanced Supersonic Technology

Spain contributes 10 percent of the funding for the Meteor Missile Program (see "European Union" in this appendix).[162]

Hypersonic R&D

Spain is not heavily involved in hypersonic aerodynamic or propulsion technology development, particularly for defense and missile applications. The largest defense contractor in Spain, Seppen, is responsible for the guidance, navigation, and control systems and the flight management system of the IXV Reentry Demonstrator. Spain additionally has a 5.35 percent share of Airbus Group (formerly known as EADS).[163]

Research and Testing Facilities

Spain does have a rocket test facility at El Arenosillo, which is the primary test site for the Spanish Ministry of Defense.[164] It specializes

---

[161] Nuclear Threat Initiative, "South Korea," web page, April 2016a.

[162] Robert Hewson, "MBDA Meteor Enters Production as Development Concludes," *Jane's Missiles & Rockets*, August 2, 2012a.

[163] Craig Caffrey, "Briefing: BAE Systems and EADS Merger," *Jane's Defence Weekly*, October 9, 2012.

[164] Instituto Nacional de Técnica Aeroespacial, "New Features: El Arenosillo," web page, undated.

in launching sounding rockets, atmospheric research, UAVs, and solar energy systems.[165]

Our literature review revealed no hypersonic wind tunnel facilities.

Sweden

Current Advanced Supersonic Technology

Sweden is a partner country in the MBDA Meteor Missile Program (see "European Union" earlier in this appendix) and is the first to integrate the missile into its air force on the Saab Gripen strike fighter.[166] Beginning in 2016, the Mach 4 air-to-air missile reached initial operating capability on the Gripen.[167]

Hypersonic R&D

Sweden developed an interest in hypersonic technology in the early 2000s but has not made significant progress in the field since 2005. In 2002, it began a three-year joint venture with Germany to develop hypersonic missiles. The collaborative effort included Bofors Dynamics Group (a defense subsidiary of Saab based in Sweden that focuses on missile systems and anti-tank weapons) as the Swedish team and was commissioned by the Swedish Office of Armaments and Defense.[168] The program was called the Hypersonic Technology Joint Program (LFK). The effort yielded a rocket-powered missile capable of operating in the Mach 4–8 range while staying below an altitude of 300 m. In 2002, the first test flight was successful and reached a speed of Mach 6.5. In 2003, the second test flight set another world record at Mach 7.[169] However, shortly thereafter, Germany withdrew from the program due to a lack of funding. While the Swedish team tried to

---

165 Instituto Nacional de Técnica Aeroespacial, undated.

166 Hewson, 2012a.

167 Tomkins, 2016.

168 Taverna, 2003. Robert Wall and Michael Taverna, "Slowing Speed," *Aviation Week and Space Technology*, Vol. 159, No. 20, 2003.

169 Wall and Taverna, 2003.

continue development, it was unable to proceed without its German counterparts.[170]

In 2009, the Swedish Defense Research Agency and Swedish Space Corporation became partners in the Future High-Altitude High-Speed Transport (FAST20xx) project sponsored by the European Space Agency.[171] This collaborative European effort aims to provide reliable transportation at hypersonic speeds to fly long distances in short periods of time. Canceled after a three-year period, the effort cost 7.3 million Euros (of which 5.1 million Euros were contributed by the European Union) and did not result in a detailed vehicle design.[172] Sweden is not currently involved in any major R&D efforts for hypersonic vehicles.

Research and Testing Facilities

Sweden is home to the Esrange rocket test launch facility operated by the European Space Research Organization in Kiruna, Sweden.[173] While the Swedish government hoped to build a spaceport at Esrange in 2009 as a part of the FAST20xx project, the primary focus of the facility is atmospheric research.[174]

Taiwan

Current Advanced Supersonic Technology

Taiwan is currently developing an indigenous ramjet engine intended to fit on the Sky Bow 3, an indigenous surface-to-air antiballistic missile that builds on previous generations that first entered into service in

---

[170] Wall and Taverna, 2003.

[171] Space Engineering & Technology, "Facts and Figures," web page, October 2, 2012a; Space Engineering & Technology, "FAST20XX (Future High-Altitude High-Speed Transport 20XX)," web page, October 2, 2012b.

[172] Space Engineering & Technology, 2012a.

[173] The Swedish Space Corporation, "Esrange Space Center," web page, undated.

[174] The Swedish Space Corporation, undated.

the 1980s.[175] The missile and engine are both still currently undergoing testing but are predicted to enter into service with the air force and navy in 2017.[176]

Hypersonic R&D

Taiwan does not appear to have an indigenous hypersonic R&D program. Most work is done at the university level using computational models for scramjet design and is based out of National Cheng Kung University and National Taiwan University.[177] However, each university lacks hypersonic wind tunnel facilities.

Research and Testing Facilities

While universities in Taiwan possess subsonic and supersonic wind tunnels, we have seen no evidence that Taiwan has hypersonic wind tunnel testing facilities.

United Kingdom

The British exit from the European Union (also known as "Brexit") in the next couple of years will require Britain to renegotiate many of its existing agreements with the European Union, including treaties on trade and technology-sharing. Prime Minister Theresa May's announcement that the UK intends to leave the common European market suggests that many of its economic ties with the European continent will change dramatically. As a result, British participation in

---

175 "Ramjet Engine for Sky Bow Sam," *Jane's Defence Weekly*, July 14, 1990.

176 Navy Recognition, "Taiwan's NCSIST Successfully Tested a Ship-Based Variant of Tien Kung III BMD Interceptor," web page, January 2, 2017.

177 Chih-Yung Wen, Yen-Sheng Chen, and Jiiun-Yue Chen, "Numerical Simulation of Complex Internal Viscous Flow in the HyShot-II Scramjet Model," 17th AIAA International Space Planes and Hypersonic Systems and Technologies Conference, April 2011; Yung-Tai Chou, Ming-Chiou Shen, Sheam-Shyun Lin, and Bor-Jang Tsai, "Flight Simulation of Hypersonic Waverider with Finlets Under Various Angles-of-Attack," 32nd Joint Propulsion Conference and Exhibit, Joint Propulsion Conferences, American Institute of Aeronautics and Astronautics, July 1996.

the European projects described here are subject to change and could decrease in the coming years as the UK further distances itself from European Union projects.

Current Advanced Supersonic Technology

The UK is the primary funder and developer of the MBDA Meteor Missile Program (see "European Union" earlier in this appendix), which entered production in 2012 and achieved initial operating capacity on the Saab Gripen in 2016.[178] Reports indicate that it will enter service with Britain's Eurofighter Typhoon in 2018.[179]

The UK also collaborated in 2002 with the United States in the development of a supersonic strike weapon—the Stand-off High-Speed Option for Counter-Proliferation (SHOC)—which would enable a fast strike against adversaries seeking to develop WMDs.[180] This project, however, appears to have been abandoned by the UK due to priorities and by the United States in favor of prompt global strike.

Hypersonic R&D

While the UK has some stake in defense contractors engaged in hypersonics research for the European Union, it has not been—and is not—a priority for the British government. After some exploration in extended-range air-to-surface needs for Future Offensive Air Systems (FOAS), including hypersonic missiles, a strategic analysis of hypersonic scramjet research stated in 2006 that, "Without a defined military requirement for HSW (high-speed weapons), the lack of UK capability in key areas and the pressures on R&D funding, it is recommended that the UK should not pursue an indigenous capability in HSW."[181] A subsequent strategic review reaffirmed this view when it described the expectation that, by 2035, the UK will have been overtaken in

---

[178] Tomkins, 2016.

[179] Hoyle, 2015.

[180] Douglas Barrie and Robert Wall, "U.K., Pentagon to Team on Supersonic Missile," *Aviation Week & Space Technology*, Vol. 157, No. 23, December 2, 2002, p. 32.

[181] "Defence Technology Strategy for the Demands of the 21st Century," Great Britain Ministry of Defence, 2016.

some technologies and may need to become accustomed to being overmatched. It specifically mentioned missile development, stating that, "The majority of missiles (including anti-ship cruise missiles) will operate at supersonic or even hypersonic speeds."[182] As a result, the British government has invested very few resources into hypersonic programs, and almost all of the research has been in collaboration with either the United States or the European Union.

Reaction Engines, a British defense contractor, is involved in two projects that develop hypersonic technologies: the SABRE engine and the LAPCAT II commercial vehicle. The SABRE engine—a hybrid engine capable of operating in both closed cycle rocket and air-breathing supersonic modes, designed for hypersonic flight into space and orbit—was developed in 2015 by Reaction Engines with BAE Systems.[183] The engine is intended to power the Skylon Space Plane—a reusable single-stage-to-orbit spacecraft that could reach speeds of up to Mach 5 (air-breathing) and Mach 25 (rocket-powered).[184] With a government award of £60 million and total investment much greater, Reaction Engines hopes to sell the commercial aircraft for $1 billion each.[185] Developers expect to have the engine in full-rig ground tests by 2019.[186]

The other project, the LAPCAT II, would be an atmospheric vehicle design for commercial transportation. Using a derivative of the SABRE (called Scimitar), Reaction Engines is working with AEA Technology and Bristol, Kingston, and York Universities to develop the early research needed for a hypersonic aircraft.[187]

---

[182] Richard Scott, "Known Unknowns: Future Operating Environment 2035," *Jane's Defence Weekly*, January 5, 2016.

[183] Kyle Maxey, "UK Government Invests £60M in Spaceplane Engine," ENGINEERING. com, July 18, 2013.

[184] Jonathan Amos, "Skylon Spaceplane Engine Concept Achieves Key Milestone," *BBCNews*, November 28, 2012.

[185] Dan Thisdell, "SABRE Cutting Path to Live Testing," *Flight International*, 2015.

[186] Thisdell, 2015.

[187] Scott, 2016.

The British company Qinetiq is working with the Australian government on the HyShot 3 project, and launched a successful test of the scramjet propulsion system in 2006.[188] The test fed into the Sustained Hypersonic Flight Experiment (SHYFE), which was scheduled to test in August 2009 but was called off after the British government declared it unnecessary.[189]

The UK owns 22.5 percent of Airbus Group (formerly EADS) and a golden share of BAE systems (giving it veto status over any change in ownership).[190] In 2012, EADS (now Airbus) was collaborating with Russian researchers on "an air-breathing technology that promises efficient operations from subsonic to hypersonic speeds of up to Mach 5."[191] EADS matched a grant of 1.7 million Euros from Russia's Skolkovo Foundation for the program in 2012.[192]

Research and Testing Facilities

The UK is home to ten hypersonic wind tunnel facilities, housed in various locations and universities across the country.[193] Some are managed by the National Wind Tunnel Facility, which attempts to open tunnel access for external use.[194] In 2016, the UK Engineering and Physical Research Council and the UK Aerodynamics Centre pledged £13.3 million to upgrade wind tunnel facilities at Imperial College London, City University London, and the Universities of Cambridge, Glasgow, Oxford, Southampton, and Cranfield.[195]

---

188 Douglas Barrie, "Speed Merchants," *Aviation Week & Space Technology*, Vol. 164, No. 14, April 3, 2006, p. 32.

189 Douglas Barrie, "Hi or LO: U.K. Hedges on High Speed Versus Stealth," *Aviation Week & Space Technology*, Vol. 167, No. 3, July 16, 2007, p. 42; Douglas Barrie, "No-Fly Zone," *Aviation Week & Space Technology*, Vol. 169, No. 8, September 1, 2008.

190 Guy Anderson, "Roxel Ramjet Test Facilities to Transfer to MBDA," *Jane's Defence Industry*, 2013.

191 Dan Thisdell, "EADS Puts Faith in Detonation Engine," *Flight International*, 2012.

192 Thisdell, 2012.

193 Aerospace Technology Institute, "UK Wind Tunnels," web page, undated-h.

194 The National Wind Tunnel Facility, "About Us," web page, undated.

195 University of Oxford, "New National Wind Tunnel Facility," web page, undated.

University of Oxford

1. *Hypersonic Gun Tunnel*: Capable of testing speeds of Mach 6, 7, and 8 up to a stagnation temperature of 1,000 K. It can also isentropically compress test gas to give "cold" test times around 300 milliseconds (ms) at 600 K.[196]
2. *High Density Tunnel*: Capable of testing speeds of Mach 4, 5, 6, and 9 with a test cross-section of 0.3 m in diameter.[197]
3. *Low Density Tunnel*: Capable of testing speeds at a range of Mach 5.5–9 using continuous hypersonic flow at low pressure. It has a test cross-section of 0.18 m in diameter.[198]
4. *T6 Free Piston Reflected Shock Tunnel*: This tunnel is still under construction, but it will be able to test at speeds of Mach 6, 7, and 8, with a total temperature up to 5,000 K, with a test flow size of 0.2–0.3 m in diameter.[199]

University of Manchester

5. *Shock Tunnel*: Capable of testing speeds of up to Mach 5. It contains a test cross-section of 0.1 m in diameter.[200]
6. *HSST Tunnel Facility*: Tests speeds of Mach 4, 5, and 7, including a supersonic/hypersonic blowdown. It has a test cross-section of 0.22 m in diameter.[201]

---

[196] Aerospace Technology Institute, "Hypersonic Gun Tunnel," web page, undated-c.

[197] Aerospace Technology Institute, "High Density Tunnel," web page, undated-a.

[198] Aerospace Technology Institute, "Low Density Tunnel," web page, undated-e.

[199] National Wind Tunnel Facility of the United Kingdom, "University of Oxford T6 Free Piston Reflected Shock Tunnel," web page, undated.

[200] Aerospace Technology Institute, "Shock Tunnel," web page, undated-g.

[201] Aerospace Technology Institute, "HSST Tunnel Facility," web page, undated-b.

Imperial College London

7. *Hypersonic Gun Tunnel*: Can test speeds of up to Mach 9 in the study of transition and turbulent boundary flows. It contains a test cross-section of 0.457 m in diameter.

Cranfield University

8. *Hypersonic Gun Tunnel*: Tests speeds of Mach 8.2–12 with intermittent blowdown and a maximum stagnation temperature of 1,290 K. It contains a test cross-section diameter of 0.214 m.[202]

Aircraft Research Association

9. *M7T Blowdown Wind Tunnel*: Tests speeds of up to Mach 8 with a test cross-section diameter of 0.31 m.[203]

Gas Dynamics, Ltd.

10. *Light Piston Isentropic Compression Facility*: Tests speeds of Mach 6.84 and 9.4 with a test cross-section diameter of 0.21 m.[204]

University of Glasgow

11. *Trisonic Tunnel:* Capable of testing speeds of up to Mach 5 using a test cross-section diameter of 0.2 m.
12. *Hypersonic Shock Tunnel*: Capable of testing speeds of up to Mach 10 for 20–80 ms.[205]

---

[202] Aerospace Technology Institute, undated-c.

[203] Aerospace Technology Institute, "M7T Blowdown Wind Tunnel," web page, undated-f.

[204] Aerospace Technology Institute, "Light Piston Isentropic Compression Facility," web page, undated-d.

[205] University of Glasgow, "Access State-of-the-Art Facilities for Aerospace Research," web page, June 2, 2016; University of Glasgow, Aerospace Sciences, "Wind Tunnel Facilities," web page, undated.

APPENDIX C

Technical and Economic Barriers to Hypersonic Systems Development

In this appendix, we summarize the principal technical and economic barriers to developing systems designed for sustained hypersonic flight.

Technical Barriers

As described earlier, numerous systems have flown hypersonically, including reentering manned capsules, ICBM RVs, and test vehicles like the X-15 and the X-51. Developing these systems were extremely complex and expensive endeavors of which few nations were capable. Developing hypersonic flight systems capable of sustained flight, such as HGVs or HCMs, is even more difficult. Both of these types of systems will be designed for one-time uses. This makes them steppingstones to the more challenging designs of reusable systems that have longer flight times and much longer operational lives. After many years of concentrated effort and investment, the United States, Russia, and China may be closing in on these one-time-use, expendable capabilities. It is unclear whether France and India will be able to achieve the same capability independently, but it is unlikely that many others will have the means or will to do the same without significant assistance from these frontrunners.

This appendix provides summary descriptions of the technology areas that are the most difficult to develop, and therefore the areas for which other nations would likely need foreign assistance in order to develop hypersonic weapon systems.

Before discussing specific technologies, a brief discussion of the hypersonic flight envelope may be helpful. By convention, the envelope starts at Mach 5 and extends to above Mach 25, which corresponds approximately to the speed that objects will sustain in low-Earth orbit. However, most air-breathing hypersonic flight concepts discussed in the literature envision flight at single-digit or "teen" Mach numbers, while HGVs will have initial speeds around Mach 20. Many physical phenomena that create design opportunities and challenges for hypersonic flight vehicles, such as lift, drag, stagnation pressure, and stagnation temperature, are functions of the vehicle speed squared (or Mach number squared), as we discussed in Appendix A. Therefore, although most of the challenges discussed in this appendix are very difficult to overcome at Mach 5, they will generally become more difficult at a rate of the square of the Mach number as designers attempt to build even higher speed vehicles.

We have organized these technical barriers into four groups:

- thermal management and materials
- air vehicle and flight control
- propulsion for hypersonic cruise missiles
- hypersonic regime testing, modeling, and simulation.

The discussions below are high-level summaries of these technical challenges designed to provide a helpful intuitive background to the reader; they are not intended to present rigorous technical foundations of hypersonic flight.

Thermal Management and Materials

The classes of hypersonic and other high-speed vehicles introduced in this report experience different environments that strongly influence their designs. Satellites operate in near vacuum conditions and consequently do not experience the intense heating rates and pressure loads caused by atmospheric gases. Ballistic RVs are subjected to extremely high aerodynamic heating rates and pressures resulting from entering the atmosphere at high speeds and potentially steep angles. The peak instantaneous heating rates of RVs can be several times higher than those experienced by HGVs and HCMs. However they occur over a significantly shorter period of time, on the order of tens of seconds, versus many minutes for HGVs and HCMs.[1] In general, the total amount of heat absorbed by RVs during their overall trajectory is lower than would be experienced by HGVs or HCMs, as described in Appendix A. Manned RVs, e.g., Space Shuttle or Apollo, also experience intense heating rates. However their larger size, aerodynamic properties, and trajectories (slower reentry, large radius nose) result in their associated heating rates being lower than those of HGVs, as discussed in Appendix A.

Thus, the compact size and higher aerodynamic heating associated with HGVs and HCMs make it more difficult to maintain their structure and internal components below their upper temperature limits. These vehicles may also tend to bend or warp because of thermal gradients across their thin structures. Also challenging is the potential for ablation, erosion, and oxidation of sharp airfoil leading edges. Creative approaches to materials, active and passive cooling, and manufacturing will be needed to address the requirement for high temperature, lightweight, affordable structures with integrated thermal barriers for these expendable vehicles.

HCMs have yet another major thermal challenge that is created by the air-breathing propulsion system. Slowing the airflow into the engine from free stream hypersonic to supersonic velocities and then burning fuel will create extreme heating loads on the engine and nozzle structures. Mitigating the effects of these high heating loads requires advances in structures, materials, and active cooling, as discussed further in the Propulsion for Hypersonic Cruise Missiles section, below.

The high temperature environments will also create challenges for HCMs' and HGVs' sensor and communications systems. For example, electro-optical sensors, radomes, and antennae will need to survive these high temperature environments and permit signals to pass with-

---

1 Reentry time for an RV is affected by a number of parameters, including reentry speed and angle and vehicle design, e.g., drag coefficient, dimensions, and weight.

out distortion, or at least with known and correctable distortions, so that the sensors and communications equipment can function. Additionally, sensors may be challenged by the ionized flow created in the higher Mach numbers of the hypersonic regime.

Air Vehicle and Flight Control

Another area of serious challenges associated with maneuverable hypersonic vehicles operating in the atmosphere is the integrated air vehicle and its flight control system. Air vehicles that fly at hypersonic speeds have very thin boundary layers. The interactions between the vehicle's shock waves, its boundary layers, and the associated laminar and turbulent flows are extraordinarily complex. The entire surface and precise shape of these flight vehicle systems are important to their aerodynamic performance.

The space shuttle had large control surfaces and multiple attitude control engines designed to help control the vehicle from reentry to landing. Additionally, the relative change in geometry from structural bending was designed to be negligible, especially relative to its characteristic size, in order to minimize the strain on the thermal tiles interfaced with the structure to ensure they remain attached to the vehicle.

On the other hand, the HGV's and HCM's smaller sizes, thinner structures, and low weights can cause them to bend and flex, which can alter their aerodynamic properties. Additionally, any shape change from material ablation or erosion from high temperatures and velocities can also change the aerodynamic characteristics of the vehicle. Consequently, adaptive flight controls that can sense changes in aerodynamic properties and adapt flight control inputs might be needed to control these vehicles in flight.

We note that the accuracy of HGVs and HCMs with conventional missions will also likely need to be higher than those of RVs due to their prospective mission requirements. Therefore, their tolerance to aerodynamic induced trajectory errors will be smaller.

Propulsion for Hypersonic Cruise Missiles

The major enabling technology for HCMs is the air-breathing propulsion system. This is also the key technical difference between HCMs and HGVs, and the attribute that gives the HCM's suite of technologies the potential for growth to other hypersonic flight systems and missions. The missile first needs to be accelerated to a speed of about Mach 5 before scramjet (supersonic combustion ramjet) operation is initiated. Scramjets are air-breathing engines (they use oxygen from the atmosphere as the oxidizer for combustion) that operate in the hypersonic regime. Allowing the flow through the engine to remain supersonic precludes the extremely high stagnation pressures, temperatures, and some of the disassociation (molecules breaking apart due to extreme heating) that would result from slowing the flow through the engine to subsonic speeds, as is done in their lower-speed cousins, ramjet engines. However, scramjets remain a challenge to design and operate and have only recently been able to produce positive thrust in flight tests.[2] Different versions of these engines are currently being tested, and as of the date of this report, developing and manufacturing a reliable production version remain an aspiration of multiple countries.

Engineers have not been able to find a propulsion cycle that will provide thrust over the entire flight envelope: subsonic through hypersonic. Combined cycle engines (CCEs) are systems that integrate two or more types of propulsion cycles, normally to provide a broader range of capability, such as a broader span of Mach number. For a notional example, a CCE might incorporate a solid rocket booster propellant into a missile's variable geometry combustion chamber. This design would have the rocket accelerate the missile from a standstill on a ground launcher to Mach 2, where ramjet operation would take over and accelerate the missile to about Mach 4.8, at which point the geometry of the inlet, combustion chamber, and nozzle would change, establishing supersonic flow through the combustion chamber and thus scramjet operation. The scramjet might accelerate the missile and continue to sustain its cruise at Mach 6. A small expendable jet engine could be substituted for the rocket–ramjet combination to form a different CCE. Neither of these notional CCEs would be simple to design

---

2 A good example is the USAF's X-51 test vehicle, which accelerated under scramjet power from Mach 4.8 to Mach 5.1 in 2013 (see Mike Wall, "Air Force's X-51A Hypersonic Scramjet Makes Record-Breaking Final Flight," Space.com, May 3, 2013).

or manufacture. If commercial reusable hypersonic systems are developed, they will employ CCEs or will need to have separate propulsion systems for different Mach regimes.

Hydrocarbon fuels[3] are being developed for use in hypersonic propulsion systems. In some cases existing fuels have been used; in other cases specialty fuels have been designed and tested. By using hydrocarbon fuels, existing infrastructures can be employed, the missiles are more easily stored at ambient temperatures, and handling the fuels may be safer than handling hydrogen, although some specialty fuels are quite toxic and dangerous. However, complex hydrocarbons will not burn readily in the extremely short timelines associated with flows through scramjets, so they must be broken down into very simple hydrocarbons and hydrogen. This process of chemically breaking down the complex hydrocarbons is an advantageous endothermic process (the fuel absorbs energy) in that the fuel is used to cool the scramjet combustor's walls. Thus, while keeping the combustor walls cool enough to remain structurally sound, energy is transferred to the fuel, the fuel is broken down into simple molecules that can be more easily burned, and the energy is returned to the combustion process rather than being lost. However, the desired endothermic chemical decomposition is not assured and could occur in ways that are not helpful to the engine's operation. One approach to encourage the desired chemical decomposition is through the use of catalysts in the fuel passages in the combustor walls.

Even with these hydrocarbon fuels broken down into very simple components like hydrogen, methane, ethane, and others, they must be injected, atomized, mixed, and burned in a supersonic flow moving through a tube (combustor) in a period of a couple thousandths of a second. In traditional jet engines, metal devices protrude from the walls of the combustion chamber to inject the fuel and to provide local areas of flow stagnation where the air and fuel can mix and a flame can reside and act as a torch igniter for the passing fuel-air mixture. None of this is possible in a scramjet, as these intrusive devices would cause

---

[3] Gasoline, diesel fuel, and aviation fuel are all forms of hydrocarbon fuels with which we are familiar.

shock waves that would result in significant stagnation pressure losses and thus lower the efficiency of the engine, potentially precluding positive thrust. The physical integrity of the penetrating injector-flame holder would also be difficult to maintain in the hot, high-speed flow stream. Therefore, this combustion process must happen extremely rapidly and quite thoroughly if positive thrust is to be created, yet without intrusive fuel injection and flame stabilization aids.

The design and operation of a scramjet engine are further complicated by additional chemical complexity called "disassociation." Also called, "real gas effects," the process of disassociation is the breaking apart of molecules due to very high temperatures. This endothermic process reduces the potential of the engine's flow stream to produce thrust. According to Heiser, these real gas effects begin around 3,000 degrees Rankine (1,700 K).

At higher Mach numbers, possibly starting around Mach 8, hydrocarbon fuels may not provide sufficient cooling capacity and designs may need to shift to using hydrogen as the fuel.

Hypersonic Regime Testing, Modeling, and Simulation

As discussed in the previous sections, the endo-atmospheric hypersonic environment is challenging not only to attain but also to operate in. Reaching and maintaining hypersonic speeds within the atmosphere requires significant capabilities involving advanced air-breathing propulsion or, in the case of the HGV, missile technology similar to what is used to perform space launches. As a result, testing hypersonic systems or subsystems also requires similar capabilities unless the testing is done on the ground. However, hypersonic wind tunnels capable of producing flight-representative hypersonic flow for extended periods (several seconds or longer) with flight equivalent stagnation temperatures and pressures are extraordinarily difficult and expensive to build. Despite even the best efforts, ground test facilities do not perfectly represent the flight environment. The test sections are small, causing the test articles to be subscale. Furthermore, the airflow is often contaminated with particulate matter, ionization, or excess water as a result of various means of heating the flow to flight-representative temperatures; test durations are short; and the "noise" or turbulence levels in the test section flows are typically significantly higher than would be found in flight.

This does not mean that high-quality hypersonic ground test facilities are not useful. Through careful examination and calibration, especially based on actual flight test data, ground test facilities can be useful and their results can be interpreted appropriately. Furthermore, wind tunnel and test instrumentation technologies continue to advance.

Integrated aero-thermo-structural computational models used to emulate hypersonic flows around flight vehicles have improved substantially with the availability of supercomputers, advancement of computational fluid dynamics, and high fidelity thermal and structural modeling. However, these computer models still do not have sufficient fidelity or accuracy required to design a hypersonic vehicle without complementary ground and flight test data. They will become more capable as computational capabilities advance and as more ground and flight test data become available. However, ground and flight-testing of vehicle designs will continue to be required for successful developments for the foreseeable future.

Economic Challenges

In addition to the technical barriers, there are formidable economic barriers to hypersonic programs. The R&D and infrastructure described earlier in this appendix for ground testing are extraordinarily expensive, as is any viable flight test program that would prepare a weapon system or a commercial system for operational use. Apart from the technical barriers, these economic barriers may be sufficient to prevent most nations from developing hypersonic weapons without foreign assistance.

As described earlier, a key challenge for nonproliferation regimes is the "dual-use" claim that the technologies will be used for commercial purposes, thus making them eligible for proliferation. However, the economic barriers to developing commercial hypersonic flight system applications make them very questionable. Possibly the most frequently cited commercial application is hypersonic airliners. Yet these would require development programs costing many billions of dollars and lasting decades. While the concept of rapid transoceanic flight is intuitively exciting, an objective cost-benefit analysis will likely make such systems bad business cases for the foreseeable future. History demonstrated that the business case for the Mach 2 Concorde was at best marginal, and the factors that worked against that business case would be multiplied for hypersonic airliners. As with the Concorde, the high research, development, procurement, infrastructure, and operating costs of a hypersonic airliner will likely make its business case unattractive. On the other hand, an air-breathing first- or second-stage space launch system may be economically viable.

Summary of Challenges

The persistent high speed and long atmospheric flight time of hypersonic vehicles result in an extremely severe operating environment requiring advanced new systems, components, materials, design tools, and test facilities. Some of the major challenges are summarized in this appendix, and the major technologies needed to address these challenges are discussed in Appendix D. Restricting the export of these technologies would impede the proliferation of hypersonic weapons because most countries would not be able to develop them without external assistance. Another major obstacle to developing a hypersonic weapon is the integration of all the enabling technologies into a working system. Therefore, the highest priority in export control is a fully integrated weapon, followed by complete subsystems, e.g., a full scramjet, a complete thermally protected structure, and a full flight control system.

To these technical barriers one must add the economic challenges of hypersonic projects. The combination of all these impediments will add potential effectiveness to a hypersonic technology proliferation control policy.

APPENDIX D

Suggested Export Control List for Hypersonic Technologies

This appendix contains ten suggested hypersonic items (plus subitems) that should be subject to export controls if a policy were established to hinder hypersonic missile proliferation. These items were identified through report reviews, meetings with subject-matter experts, and the authors' own assessments. They are considered to be the key enablers for HGVs and HCMs and require significant technical know-how, time, and resources to develop. Most countries would need to either import them or obtain substantial external assistance to develop them. As such, controlling the export of these items should help impede the proliferation of hypersonic weapons. As we have discussed previously in the report, many hypersonic technologies are dual use, i.e., they can support civilian applications such as hypersonic airliners. Therefore, we propose a two-tier control system to allow the carefully controlled export of the technologies intended for civilian uses. We define these items using the MTCR format and identify how they might fit into the existing MTCR Annex.

Standard Additions to Export Controls

Some details are not included because, given current international practice, they would be standard additions in any export control policy. Under the MTCR and other export control regimes it is now standard practice to add details to the descriptions of controlled items by adding such expansions as "production facilities," "production equipment," "software," "technical data," and "technology." The suggestions in this appendix represent the kernels of controlled items; it is here assumed that the expansions would be added as appropriate to a control list formulated within or outside of the MTCR.

In addition, the MTCR and other export control regimes include a "catch-all" rule and a "no-undercut" rule.[1] The catch-all rule establishes export controls for items for WMD delivery systems, even if such items are not on a control list. The no-undercut rule, when applied to an export review of an item on the control list, requires consultation with a partner who has denied the export of that item and has notified the other partners of the denial. This appendix assumes that both rules would be applied to exports within or outside of the MTCR.

Specific Suggestions for Export Controls

The following ten items are divided into three that are proposed for a strong presumption of export denial, because they have no credible dual-use potential, and seven that are proposed for case-by-case review because of their dual-use potential for such applications as civilian aircraft. Under the MTCR, the first three items would be characterized as Category I and the last seven as Category II.

Category I List

The proposed Category I list consists of three items that would be subject to a strong presumption of export denial.

- **Complete Delivery Vehicle:** Glide vehicle with any range capable of Mach 5+, or UAV[2] with greater than or equal to a 300-km range and capable of Mach 5+ flight, with or without payload capability. The range of a glide vehicle is dependent upon its ini-

[1] For more information, see FAQs No. 12 and 14 on the MTCR website (MTCR, "Frequently Asked Questions (FAQs)," web page, undated-a).

[2] *UAVs* include missiles such as HCMs.

tial release conditions, i.e., altitude, speed, and flight path angle; therefore, it would not be practical (or indeed possible) to control the export of HGVs based on their range. Additionally, both HGVs and HCMs can potentially be used without an additional payload; i.e., their high kinetic energy can be used to destroy targets. The 300-km range capability is applied to HCMs because this quantity is standard in the MTCR and, indeed, has substantial international acceptance. The Category I control of complete delivery vehicles should apply not only to assembled vehicles but also to shipments of enough major subsystems to effectively provide access to a complete delivery vehicle. (New to MTCR Annex item 1.A)

- **HGVs in addition to RVs.** HGV trajectories can be designed to be completely within the atmosphere. In other words, HGVs may not exit the atmosphere and thus would not reenter the atmosphere. Therefore, HGVs are not necessarily RVs. Furthermore, HGVs have trajectories and maneuverability characteristics that are significantly different than those traditionally associated with RVs. Moreover, the MTCR only controls RVs as Category I if they meet the criteria of a 500-kg payload and a greater than 300-km range and are not designed as a peaceful payload. Note that rocket boosters for HGVs are generally controlled as Category I in the current MTCR Annex. (New to MTCR Annex item 2.A.1.b)
- **Warheads for Mach 5+**
  - Conformal warheads
  - Safing, arming, fusing, and firing systems
  - Weapon payload deployment and dispensing mechanisms designed for hypersonic vehicles (applies to endo-atmospheric submunitions).

Warheads for HGVs and HCMs may need to be especially designed to fit within the thin structure of these vehicles. HGVs and HCMs can also rely on their kinetic energy alone to destroy their targets. The safing, arming, fusing, and firing systems of HGVs and HCMs will also need to be specifically designed to withstand and operate in hypersonic flight conditions. Payload deployment and dispensing mechanisms designed to operate at Mach 2 and above should be controlled because they enable the use of submunitions on hypersonic vehicles. (New to MTCR Annex item 2.A.1.f)

Category II List

The proposed Category 2 list includes dual-use items that are assessed for export control on a case-by-case review.

- **Scramjet and combined cycle engines capable of operation above Mach 5**, including
  - inlet starting and flow control systems and techniques
  - injector, flame-holder design and control
  - fuel-air mixing and combustion enhancement, including assisted combustion, e.g., plasma and others; fuel piloting; cold-start approaches
  - sensors used in scramjets to inform engine control
  - cooling approaches for scramjets.

Scramjets are a class of air-breathing engines that operate in the hypersonic regime. They are already subject to MTCR controls along with their (currently unspecified) "devices to regulate combustion and specially designed components." However, these engines would also power future hypersonic commercial transports, so they are dual use. The specific subsystems listed represent some of the most challenging technologies involved in the development of scramjets and are therefore included separately. Combined cycle engines, also broadly mentioned in the MTCR Annex, are integrated propulsion systems designed to provide thrust from a lower speed (possibly from a standstill on a runway) into the hypersonic regime and should, therefore, be included in this list. These engines could also power commercial transports and are therefore considered dual use. (Relates to MTCR items 3.A.2)

- **Hydrocarbon fuels unique for Mach 5+ sustained flight, fuel-catalyzing approaches**

  By using hydrocarbon fuels, existing infrastructures can be employed, the missiles are more easily stored at ambient temperatures, and handling the fuels is safer than handling hydrogen. Hydrocarbon fuels that are being developed specifically for use in hypersonic propulsion systems should be considered for the control list as a Category II, because they may also support commercial hypersonic transports. However, any other hydrocarbon fuels that are also used widely by commercial or military applications should not be controlled. When these complex hydrocarbon fuels are used to regeneratively cool engine structures, catalysts are used to break down the fuels into simple hydrocarbons and hydrogen, enhancing heat absorption and preparing the fuel for subsequent combustion. These catalysts and catalyzing approaches should be considered for the control list as Category II, because they may also support commercial hypersonic transports. (Relates to MTCR Annex item 4.C.2)
- **Materials and thermal protection for Mach 5+ sustained flight, including**
  - high-temperature ceramics, carbon-carbon, and protective coatings
  - lightweight thermal protection for airframes, including shape-stable leading edges, ultra-high temperature ceramics, and ablative materials.

  Thermal control is a major challenge associated with hypersonic atmospheric flight because of the resulting high aerodynamic heating environment. The control list should include high temperature and low weight materials that enable hypersonic flight. However, their application in hypersonic commercial transport makes them Category II. (Relates to MTCR Annex items 2.A.1.b, 6.A.2, and 6.C)
- **Sensors, navigation, and communications for Mach 5+ flight, including through ionized flows**
  - apertures, radomes, and antennae

- terminal guidance for Mach 5+ weapons.

Sensors and antennae operating on vehicles traveling at the upper end of the hypersonic regime within the atmosphere are challenged by the ionized flow generated by these vehicles. Similarly, new terminal guidance systems are needed to support hypersonic weapons. These technologies could also be used to support hypersonic commercial transports and are therefore considered dual use. (Relates to MTCR Annex items 2.A.1.d, 6.C.5, and 9.A.1)

- **Flight controls for Mach 5+ vehicles**, including systems to address thermal-structural, aerodynamic, and mechanical vibrations and associated interactions. Hypersonic vehicles traveling within the atmosphere experience significant structural vibrations exacerbated by high heating rates, thermal gradients, and steady and unsteady aerodynamic loads. Flight controls specifically designed to address these interactions are needed. Forms of this technology would also be applied to commercial hypersonic transports and are therefore considered dual use. (Relates to MTCR Annex item 10)
- **Design tools and modeling of effects at Mach 5+, including**
  - integrated, system-level computational tools that have been anchored with reliable ground and flight test data
  - Mach 5+ flight test data.

High-fidelity hypersonic vehicle design tools that have been anchored with appropriate test data can help shorten the development and design cycle of these vehicles and reduce the testing requirements. Because accurate testing is difficult and extremely expensive, limiting access to tools that may help third parties reduce the need for ground- and flight-testing as they design hypersonic systems will, in turn, limit proliferation. These tools can be used to support the development of HGVs and HCMs but also of hypersonic civilian transports. (Relates to MTCR Annex item 16.D)

- **Ground simulation and testing capabilities (including diagnostic tools) for the development of Mach 5+ sustained flight vehicles**, including combustion, thermal, and vibrational conditions and effects. Ground simulation and testing capabilities associated with hypersonic vehicles are expensive and challenging. Such capabilities also reduce the need for flight tests and potentially accelerate the development of these vehicles. Limiting the access to capable ground-testing capabilities limits proliferation. These capabilities can, of course, be used to support the development of hypersonic commercial transports, so they are considered dual use. (Relates to MTCR Annex items 15.B.2, 15.B.6 and 16.D.1)

References

Acton, James M., "Silver Bullet? Asking the Right Questions About Conventional Prompt Global Strike," Carnegie Endowment for International Peace, Washington, D.C., 2013.

"Aeronautics and Astronautics: New Findings from Marche Polytechnic University in the Area of Aeronautics and Astronautics Described," *Defense & Aerospace Week*, June 12, 2013, p. 320.

"Aeronautics and Astronautics: New Aeronautics and Astronautics Study Results Reported from Amirkabir University of Technology," *Defense & Aerospace Week*, July 11, 2014a, p. 76.

"Aeronautics and Astronautics: Findings from F. Santoro et al Update Understanding of Aeronautics and Astronautics (The Italian Spacegate: Study and Innovative Approaches to Future Generation Transportation Based on High Altitude Flight)," *Defense & Aerospace Week*, July 30, 2014b, p. 159.

"Aeronautics and Astronautics: Researchers from Technical University Report on Findings in Aeronautics and Astronautics Performances of a Small Hypersonic Airplane (HyPlane)," *Defense & Aerospace Week*, September 16, 2015, p. 145.

"Aerospace Research: Findings from Seoul National University Yields New Data on Aerospace Research," *Defense & Aerospace Week*, December 11, 2013, p. 61.

"Aerospace Research: Recent Research from Babol University of Technology Highlight Findings in Aerospace Research (Heat Reduction Using Counterflowing Jet for a Nose Cone with Aerodisk in Hypersonic Flow)," *Defense & Aerospace Week*, February 25, 2015, p. 130.

"Aerospace Research: Reports from University of Tehran Advance Knowledge in Aerospace Research," *Defense & Aerospace Week*, November 7, 2012, p. 389.

"Aerospace Research: Study Results from Pusan National University in the Area of Aerospace Research Reported (Multidimensional Flux Difference Splitting Schemes)," *Defense & Aerospace Week*, September 23, 2015, p. 134.

Aerospace Technology Institute, "High Density Tunnel," web page, undated-a. As of March 17, 2017:
http://www.ati.org.uk/aerodynamics/facilities/high-density-tunnel-141/

———, "HSST Tunnel Facility," web page, undated-b. As of March 17, 2017:
http://www.ati.org.uk/aerodynamics/facilities/hsst-tunnel-facility-71/

———, "Hypersonic Gun Tunnel," web page, undated-c. As of March 17, 2017:
http://www.ati.org.uk/aerodynamics/facilities/hypersonic-gun-tunnel-140/

———, "Light Piston Isentropic Compression Facility," web page, undated-d.

———, "Low Density Tunnel," web page, undated-e. As of March 17, 2017:
http://www.ati.org.uk/aerodynamics/facilities/low-density-tunnel-142/

———, "M7T Blowdown Wind Tunnel," web page, undated-f. As of March 17, 2017:
http://www.ati.org.uk/aerodynamics/facilities/m7t-blowdown-wind-tunnel-3/

———, "Shock Tunnel," web page, undated-g. As of March 17, 2017:
http://www.ati.org.uk/aerodynamics/facilities/shock-tunnel-68/

———, "UK Wind Tunnels," web page, undated-h. As of March 17, 2017:
http://www.ati.org.uk/technology/wind-tunnels/

Aerospace Testing Facilities Available in India, "Title of the Aerospace Test: 0.25m Hypersonic Wind Tunnel," web page, undated-a. As of March 17, 2017:
http://atfi.dlis.du.ac.in/aerodemo.php?aeroid=221&noimage.jpg=View

———, "Title of the Aerospace Test: 0.3m Hypersonic Wind Tunnel," web page, undated-b. As of March 17, 2017:
http://atfi.dlis.du.ac.in/aerodemo.php?aeroid=219&noimage.jpg=View

———, "Title of the Aerospace Test: 0.5m Hypersonic Wind Tunnel," web page, undated-c. As of March 17, 2017:
http://atfi.dlis.du.ac.in/aerodemo.php?aeroid=220&noimage.jpg=View

———, "Title of the Aerospace Test: 0.5m Hypersonic Shock Tunnels HST2," web page, undated-d. As of March 17, 2017: http://atfi.dlis.du.ac.in/aerodemo.php?aeroid=215&noimage.jpg=View

———, "Title of the Aerospace Test: 1 m Hypersonic Wind Tunnel," web page, undated-e. As of March 17, 2017:
http://atfi.dlis.du.ac.in/aerodemo.php?aeroid=223&noimage.jpg=View

———, "Title of the Aerospace Test: Hypersonic Shock Tunnel," web page, undated-f. As of March 17, 2017:
http://atfi.dlis.du.ac.in/aerodemo.php?aeroid=222&noimage.jpg=View

———, "Title of the Aerospace Test: Free Piston Driven Hypersonic Shock Tunnel HST3," web page, undated-g. As of March 17, 2017:
http://atfi.dlis.du.ac.in/aerodemo.php?aeroid=216&noimage.jpg=View

———, "Title of the Aerospace Test: Hypersonic Shock Tunnel HST4," web page, undated-h. As of March 17, 2017: http://atfi.dlis.du.ac.in/aerodemo.php?aeroid=217&noimage.jpg=View

———, "Title of the Aerospace Test: Hypersonic Shock Tunnel HST5," web page, undated-i. As of March 17, 2017: http://atfi.dlis.du.ac.in/aerodemo.php?aeroid=218&noimage.jpg=View

"AGM-88E Advanced Anti-Radiation Guided Missile (AARGM)," *Jane's Air-Launched Weapons*, March 1, 2017. As of March 16, 2017: https://janes.ihs.com/Janes/DisplayFile/JALW3723

"Alkenes: Report Summarizes Alkenes Study Findings from Q.L. Liu and Co-Researchers (Ethylene Flame Dynamics and Inlet Unstart in a Model Scramjet)," *Defense & Aerospace Week*, December 10, 2014, p. 122.

Al-Rodhan, Nayef, "Hypersonic Missiles and Global Security," *The Diplomat*, November 13, 2012. As of March 20, 2017: http://thediplomat.com/2015/11/hypersonic-missiles-and-global-security/

American Institute of Aeronautics and Astronautics, "21st AIAA International Space Planes and Hypersonic Systems and Technology Conference (Hypersonics 2017)," web page, undated-a. As of May 17, 2017: https://www.aiaa.org/EventDetail.aspx?id=33740

———, "21st International Space Plane and Hypersonic Systems and Technology Conference," web page, undated-b. As of May 17, 2017: http://hypersonic2017.xmu.edu.cn

Amos, Jonathan, "Skylon Spaceplane Engine Concept Achieves Key Milestone," *BBCNews*, November 28, 2012. As of March 16, 2017: http://www.bbc.com/news/science-environment-20510112

Anderson, Guy, "Roxel Ramjet Test Facilities to Transfer to MBDA," *Jane's Defence Industry*, 2013.

Anderson, John David, *Hypersonic and High Temperature Gas Dynamics*, Washington, D.C.: American Institute of Aeronautics and Astronautics, 2000.

Andoya Space Center (ASC), "About Us," web page, undated-a. As of August 29, 2017: http://andoyaspace.no/?page_id=47

Andoya Space Center (ASC), "Current GCI Missions," web page, undated-b. As of August 29, 2017: http://andoyaspace.no/?page_id=2232

Anonymous, "Australia and USA in HiFire Link-Up," *Flight International*, Vol. 170, No. 5063, November 2006, p. 32.

———, "14-X Hypersonic Vehicle Details Given," *Flight International*, Vol. 179, No. 5287, April 2011, p. 19.

Arming India, "India: BrahMos Aerospace Chief Comments on Ambition to Achieve 90-Degree Steep Dive Capability," interview with Sudhir Kumar Mishra, March 27, 2016.

Aroor, Shiv, "BIG! Indian Govt Clears Decks for BrahMos Export," *Livefist Defence*, June 3, 2016. As March 20, 2017:
https://www.livefistdefence.com/2016/06/big-indian-govt-clears-decks-for-brahmos-export.html?utm_source=feedburner&utm_medium=feed&utm_campaign=Feed%253A+blogspot%252FUQMw+%2528LiveFist+-+The+Best+of+Indian+Defence%2529

Australian Government Department of Defence, "Valuable Lessons Learnt in Latest Hypersonic Flight Trials," web page, August 12, 2015. As of May 17, 2017:
https://www.dst.defence.gov.au/news/2015/08/12/valuable-lessons-learnt-latest-hypersonic-flight-trial

Barrie, Douglas, "Speed Merchants," *Aviation Week & Space Technology*, Vol. 164, No. 14, April 3, 2006, p. 32.

———, "Hi or LO: U.K. Hedges on High Speed Versus Stealth," *Aviation Week & Space Technology*, Vol. 167, No. 3, July 16, 2007, p. 42.

———, "No-Fly Zone," *Aviation Week & Space Technology*, Vol. 169, No. 8, September 1, 2008.

Barrie, Douglas, and Robert Wall, "U.K., Pentagon to Team on Supersonic Missile," *Aviation Week & Space Technology*, Vol. 157, No. 23, December 2, 2002, p. 32.

Bedi, Rahul, "DRDO to Build New Missile Test Range on Southeast Coast of India," *Jane's Defence Weekly*, Vol. 49, No. 51, November 28, 2012.

"Belgium Plans to Create Own National Space Agency," *Sputnik News*, March 15, 2016. As of March 10, 2017:
http://www.spacedaily.com/reports/Belgium_Plans_to_Create_Own_National_Space_Agency_999.html

Bertin, John J., *Hypersonic Aerothermodynamics*, Washington, D.C.: American Institute of Aeronautics and Astronautics, 1994.

"Boeing Announces Involvement in Major Australia-U.S. Hypersonics Research Project," *Defense Daily International*, Vol. 8, No. 2, January 12, 2007, p. 1.

Bouillot, J. C., "French Approach in Future Launch Systems," *Acta Astronautica*, Vol. 17, No. 8, August 1988, pp. 793–805. As of March 15, 2017:
http://www.sciencedirect.com/science/article/pii/0094576588901634

"Bruisers, Incoming: Anti-Ship Strike Takes and Eastwards Fix," *IHS Jane's Navy International*, October 16, 2013.

Butterworth-Hayes, Phillip, "Europe Speeds Up Hypersonics Research," *Aerospace America*, 2008, p. 24.

Caffrey, Craig, "Briefing: BAE Systems and EADS Merger," *Jane's Defence Weekly*, October 9, 2012.

Chou, Yung-Tai, Ming-Chiou Shen, Sheam-Shyun Lin, and Bor-Jang Tsai, "Flight Simulation of Hypersonic Waverider with Finlets Under Various Angles-of-Attack," 32nd Joint Propulsion Conference and Exhibit, Joint Propulsion Conferences, Washington, D.C.: American Institute of Aeronautics and Astronautics, July 1996. As of March 20, 2017: http://arc.aiaa.org/doi/abs/10.2514/6.1996-2892

Czek, Alex, "Avro Canada CF-105 Arrow Supersonic Interceptor Aircraft Prototype," *Military Factory*, August 31, 2016. As of March 17, 2017: http://www.militaryfactory.com/aircraft/detail.asp?aircraft_id=680

de Araújo Martos, João Felipe, Israel da Silveira Rêgo, Sergio Nicholas Pachon Laiton, Bruno Coelho Lima, Felipe Jean Costa, and Paulo Gilberto de Paula Toro, "Experimental Investigation of Brazilian 14-X B Hypersonic Scramjet Aerospace Vehicle," *International Journal of Aerospace Engineering*, Vol. 2017, No. 50, May 2, 2017. As of March 20, 2017: https://www.hindawi.com/journals/ijae/2017/5496527/

de Lima Cardoso, Ronaldo, and Marcos da Silva e Souza, "Brazilian 14-x S Hypersonic Scramjet Aerospace Vehicle Dimensional Design at Mach Number 7," 22nd International Congress of Mechanical Engineering (COBEM 2013), Ribeirão Preto, Brazil, November 2013. As of March 20, 2017: http://www.henrynagamatsu.com/wp-content/uploads/2014/05/Brazilian-14-X-S-Hypersonic-Scramjet-Aerospace-Vehicle-Dimensional-Design-at-Mach-Number-7.pdf

"Defence Technology Strategy for the Demands of the 21st Century," Great Britain Ministry of Defence, 2016. As of August 17, 2017: http://trove.nla.gov.au/work/26039462?selectedversion=NBD41248226

Delft University of Technology, "HTFD Hypersonic Wind Tunnel," web page, undated. As of August 17, 2017: https://www.tudelft.nl/lr/organisatie/afdelingen/aerodynamics-wind-energy-flight-performance-and-propulsion/facilities/htfd-hypersonic-windtunnel/

Department of Aerospace Engineering: Indian Institute of Technology Madras, "Research—Aerodynamics," web page, undated. As of March 27, 2017: http://www.ae.iitm.ac.in/files/aero.htm

Dinu, Mihai-Stefan, "Strategic Tendencies of Some Major International State Actors on 2011," *Strategic Impact*, Vol. 42, 2012, pp. 37–43.

Dumiak, Michael, "North Star," *Aviation Week & Space Technology*, Vol. 174, No. 32, September 3, 2012.

Egozi, Arie, "Israel Helps India Develop Scramjet Demonstrator," *Flight International*, July 11, 2007. As of March 20, 2017:
https://www.flightglobal.com/news/articles/israel-helps-india-develop-scramjet-demonstrator-215441/

Embassy of the Republic of Belarus in the Republic of India, "Scientific and Technical Cooperation," web page, undated. As of March 17, 2017:
http://india.mfa.gov.by/en/bilateral_relations/scientific/

ESTEC: European Space Research and Technology Centre, "About Us," web page, February 19, 2016. As of August 21, 2017:
http://www.esa.int/About_Us/ESTEC/ESTEC_European_Space_Research_and_Technology_Centre

"Europe, Japan Closer to Hypersonics Tech Plan," *Aerospace America*, May 2014, p. 5. As of March 20, 2017:
http://archive.aerospaceamerica.org/Documents/AerospaceAmerica%20PDFs%20-%202014/May%202014/AA_May2014.pdf

"European Meteor Missile Test Fired over Sweden," *Defense Update*, undated. As of March 10, 2017:
http://www.defense-update.com/newscast/0308/news/news2103_meteor.htm

Fadel, Leith, "Iran to Develop Supersonic Cruise Missile," *AMN Al Masdar News*, August 22, 2016. As of March 20, 2017:
https://www.almasdarnews.com/article/iran-develop-supersonic-cruise-missile/

Falempin, Francois, and Laurent Serre, "French Flight Testing Program LEA Status," NATO Research & Technology Organization, RTO-EN-AVT-185, undated, p. 17-5, Figure 5.

———, "French Flight Testing Program LEA Status in 2009," Washington, D.C.: American Institute of Aeronautics and Astronautics, 2011. As of March 15, 2017:
https://arc.aiaa.org/doi/abs/10.2514/6.2011-2200

Fokker, "Fokker, NLR, Airborne and TU Delft Start Maintenance Centre for Composites," press release, June 16, 2015. As of August 21, 2017:
http://www.nlr.org/news/fokker-nlr-airborne-and-tu-delft-start-maintenance-centre-for-composites/

"France Studies Nuclear Missile Replacement," *Defense News*, December 1, 2014, p. 22.

German-Dutch Wind Tunnels, "About Us," web page, undated. As of March 16, 2017:
https://www.dnw.aero/about-dnw/about-dnw/

Goodrich, Malinda, Jenele Gorham, Wm. Noel Ivey, Sarah Kim, Marieke Lewis, and Carl Minkus, "Wind Tunnels of the Eastern Hemisphere," Washington, D.C.: The Library of Congress, August 2008. As of March 15, 2017:
https://www.loc.gov/rr/frd/pdf-files/Eastern_Hemisphere_Wind_Tunnels.pdf

Grevatt, Jon, "BrahMos Outlines $2bn Export Target for Supersonic Cruise Missile," *Jane's Defence Industry*, November 8, 2007.

Gubrud, Mark, "Just Say No," *Bulletin of the Atomic Scientists*, June 25, 2015. As of May 2, 2017:
http://thebulletin.org/test-ban-hypersonic-missiles/just-say-no

Gupta, Archit, "Hypersonic Aircraft at Mach 6.5 Powered by DRDO2014," *Indian Aviation News*, October 23, 2014. As of March 17, 2017:
http://spaceflight101.com/rlv-td-hex-preview/

"Haeseong (SSM-700K)," *Jane's by IHS Markit: Weapons: Naval*, March 4, 2017. As of March 20, 2017:
https://janes.ihs.com/Janes/Display/1499689

Heiser, William H., and David T. Pratt, *Hypersonic Airbreathing Propulsion*, Washington, D.C.: American Institute of Aeronautics and Astronautics, 1994.

Hewson, Robert, "MBDA Meteor Enters Production as Development Concludes," *Jane's Missiles & Rockets*, August 2, 2012a.

———, "India's DRDO Preparing for Hypersonic Test," *Jane's Defence Weekly*, Vol. 49, No. 46, October 24, 2012b.

Hiro, Dilip, "The Most Dangerous Place on Earth," WarIsBoring.com, April 4, 2016. As of May 2, 2017:
http://warisboring.com/articles/the-most-dangerous-place-on-earth/

Hoyle, Craig, "PARIS: MBDA on Target for Meteor Introduction," *Flight Daily News*, June 16, 2015.

Hughes, Robin, "U.S. Navy Conducts First AARGM Block 1 Live-Fire Test," *Jane's Missiles & Rockets*, September 30, 2015.

IAI: Israel Aerospace Industries, "Wind Tunnels Center," web page, undated. As of March 20, 2017:
http://www.iai.co.il/2013/37113-en/Groups_EDG.aspx

IHS Jane's 360: News and Defense Headlines, "Air Launched Weapons," web page, July 30, 2006. As of March 10, 2017:
http://www.janes.com/defence/weapons/air-launched-weapons

"Indian Reusable Launch Vehicle Testbed Ready for High-Speed Flight," *Spaceflight 101*, May 21, 2016. As of March 17, 2017:
https://xissufotoday.space/2016/05/
indian-reusable-launch-vehicle-testbed-ready-for-high-speed-flight/

InfoLabel, "Engineering–Research Consultants (Miscellaneous) in Singapore," web page, undated. As of March 20, 2017:
http://local.infobel.sg/1242/engineering_research_consultants_miscellaneous-singapore/index_1.htm

Institute of Aerodynamics and Flow Technology, "Hypersonic Wind Tunnel Cologne (H2K)," web page, undated. As of March 15, 2017:
http://www.dlr.de/as/en/desktopdefault.aspx/tabid-194/407_read-5448/

Instituto Nacional de Técnica Aeroespacial, "Nuevas Funciones: El Arenosillo [New Features: El Arenosillo]," web page, undated. As of March 16, 2017:
http://www.inta.es/opencms/export/sites/default/INTA/es/quienes-somos/historia/el-arenosillo/

"Iran to Build Supersonic Maritime Cruise Missiles Soon: Dehqan," *PRESSTV News*, August 21, 2016. As of March 20, 2017:
http://www.presstv.com/Detail/2016/08/21/481020/Iran-Defense-Minister-Hossein-Dehqan-cruise-missiles-torpedoes

"Israel Military Industries Ltd (IMI)," *Jane's Space Systems and Industry*, March 7, 2007.

Israel's Navy New Anti-Ship Missile," *Defense Update News*, May 27, 2016. As of March 20, 2017:
http://defense-update.com/20160327_israeli_anti_ship_missile.html

Italian Aerospace Research Centre, "Cira and Prora," web page, undated-a. As of March 16, 2017:
http://zope.gd-a-17.vps.redomino.com/Projects/cira/en/chi-siamo-en/cira-e-prora

———, "Facilities," web page, undated-b. As of March 16, 2017:
http://zope.gd-a-17.vps.redomino.com/Projects/cira/en/impianti-en

———, "HYPROB," web page, undated-c. As of March 16, 2017:
http://zope.gd-a-17.vps.redomino.com/Projects/cira/en/settori-en/propulsione-aerospaziale-e-flussi-reagenti/programmi/propulsori-aerospaziali/Progetti/hyprob

———, "LAPCAT II: Long-Term Advanced Propulsion Concepts and Technologies," web page, undated-d. As of March 16, 2017:
http://zope.gd-a-17.vps.redomino.com/Projects/cira/en/settori-en/propulsione-aerospaziale-e-flussi-reagenti/programmi/propulsori-aerospaziali/Progetti/lapcat-ii

ITU: The International Telecommunication Union, "Biography of Lim Seng," web page, undated. As of August 29, 2017:
https://www.itu.int/en/ITU-R/space/workshops/danang-2015/Documents/Presentations/Lim%20Seng%20-%20Bio.pdf

"JAXA 2025 (JAXA Long-Term Vision)," YouTube, April 9, 2009. As of March 20, 2017:
https://www.youtube.com/watch?v=jgs8G_EScz4

Jennings, Gareth, "Meteor Trials Near Conclusion," *Jane's Missiles & Rockets*, July 14, 2011.

Jons, Lorna, "Belgium–AIM-9X-2 Sidewinder Missiles," Defense Security Cooperation Agency, Washington, D.C., September 26, 2013. As of March 10, 2017:
http://www.dsca.mil/major-arms-sales/belgium-aim-9x-2-sidewinder-missiles

Ju, Yiguang, Skip Williams, and Joanna Austin, "Propellants and Combustion," *Aerospace America*, December 2008, p. 68.

Kandebo, Stanley W., "France, Russia to Join," *Aviation Week & Space Technology*, March 26, 2001, p. 60. As of March 13, 2017:
http://aviationweek.com/awin/france-russia-join

Kemburi, Kalyan M., "High-Speed Cruise Missiles in Asia: Evolution or Revolution?" *Fair Observer*, March 19, 2014. As of March 20, 2017:
http://www.fairobserver.com/region/asia_pacific/
high-speed-cruise-missiles-asia-evolution-revolution-63152/

"Kh-41 (3M82/Moskit/P-80/P-270)," *Jane's Strategic Weapon Systems*, October 13, 2011.

Laurence, D., and Wolfgang Rodi, *Engineering Turbulence Modelling and Experiments-4*, Oxford, UK: Elsevier, April 14, 1999, p. 394.

Lewis, David, and Tom Forbes, "Researchers at University of Queensland Mothball Scramjet Experiment After Failed Test in Norway," *Australia Broadcasting Corporation News*, September 19, 2013.

Loctier, Denis, "Will Hypersonic Passenger Planes Ever Be a Reality?" *Euro News*, February 3, 2015. As of March 20, 2017:
http://www.euronews.com/2015/03/02/hypersonic-airlines

Maxey, Kyle, "UK Government Invests £60M in Spaceplane Engine," ENGINEERING.com, July 18, 2013. As of March 16, 2017:
http://www.engineering.com/DesignerEdge/DesignerEdgeArticles/
ArticleID/6023/UK-Government-Invests-60M-in-Spaceplane-Engine.aspx

Menon, Jay, "Homegrown Hypersonics," *Aviation Week & Space Technology*, Vol. 174, No. 42, November 26, 2012, p. 51. As of March 17, 2017:
http://web.a.ebscohost.com/ehost/detail/detail?sid=d088530f-6fc2-4dc0-b85f-fe03f
edf26d1%40sessionmgr4006&vid=0&hid=4104&bdata=JnNpdGU9ZWhvc3Qtb
Gl2ZQ%3d%3d#AN=84384240&db=mth

Metcalfe, Tom, "Blazing-Fast Hypersonic Jet on Track for 2018 Launch," Live Science, May 26, 2016. As of March 17, 2017:
http://www.livescience.com/54877-hypersonic-hifire-engine-test-flight.html

Middlebury Institute of International Studies at Monterey, *Nonproliferation Review*, Vol. 22, No. 2, June 2015. As of May 10, 2010:
http://www.nonproliferation.org/research/nonproliferation-review/npr-22-2/

"Military and Technical Co-Operation: BrahMos Missile Air-Based Modification To Be Ready by Late 2014," *Interfax: Russia & CIS Military Information Weekly*, March 1, 2013.

Missile Threat CSIS Missile Defense Project, "Air-Sol Moyenne Portée (ASMP/ASMP-A)," web page, November 30, 2016. As of March 10, 2017:
https://missilethreat.csis.org/missile/asmp/

Mission Technology Control Regime, "Frequently Asked Questions (FAQs)," web page, undated-a. As of August 17, 2017:
http://mtcr.info/frequently-asked-questions-faqs/

———, "MTCR Guidelines," web page, undated-b. As of August 15, 2017:
http://mtcr.info/mtcr-guidelines/

———, "MTCR Annex," web page, undated-c. As of August 15, 2017:
http://mtcr.info/mtcr-annex/

Mitsubishi Heavy Industries, "20 cm Hypersonic Wind Tunnel," web page, undated-a: As of March 20, 2017:
https://www.mhiglobal.com/products/expand/hypersonic_wind_tunnel_supply_result_01.html

———, "High Enthalpy Shock Tunnel (HIEST)," web page, undated-b. As of March 20, 2017:
https://www.mhi-global.com/products/expand/shock_wind_tunnel_supply_result_01.html

Mizokami, Kyle, "Bullseye: The 5 Most Deadly Anti-Ship Missiles of All Time," *The National Interest*, March 13, 2015. As of March 20, 2017:
http://nationalinterest.org/feature/bull's-eye-the-5-most-deadly-anti-ship-missiles-all-time-12411?page=2

Morley, T. Scott, "From Singapore to Sydney: A Prize for Hypersonic Point to Point Transportation, *International Space University*, March 25, 2008. As of March 20, 2017:
iisc.im/wp-content/uploads/2015/08/Morley_Scott_PA_2008_1.pdf

Morring, Frank, Jr., and, Michael Mecham, "'One JAXA,'" *Aviation Week & Space Technology*, Vol. 163, No. 21, November 28, 2005, p. 64.

MTCR—*See* Mission Technology Control Regime.

Nadimi, Farzin, "Iran and China Are Strengthening Their Military Ties," The Washington Institute, November 22, 2016. As of March 20, 2017:
http://www.washingtoninstitute.org/policy-analysis/view/iran-and-china-are-strengthening-their-military-ties

Nair, Kiran Krishnan, "Apples and Oranges: Why Comparing India's Reusable Launch Vehicle with the Space Shuttle Is Totally Out of Place," *The Space Review*, May 23, 2016. As of March 20, 2017:
http://www.thespacereview.com/article/2990/1

National Academies of Sciences, Engineering, and Medicine, *A Threat to America's Global Vigilance, Reach, and Power- High-Speed Maneuvering Weapons: Unclassified Summary*, Washington, D.C.: The National Academies Press, 2016. As of April 18, 2017:
http://www.nap.edu/23667/

National Wind Tunnel Facility of the United Kingdom, "About Us," web page, undated. As of March 17, 2017:
http://www.nwtf.ac.uk/html/index.html

National Wind Tunnel Facility of the United Kingdom, "University of Oxford T6 Free Piston Reflected Shock Tunnel," web page, undated. As of March 17, 2017:
http://www.nwtf.ac.uk/html/datasheets/Oxford_HS2_1.pdf

Navy Recognition, "Naval Forces News–Japan," web page, November 16, 2015. As of March 20, 2017:
http://www.navyrecognition.com/index.php/news/defence-news/2017/august-2017-navy-naval-forces-defense-industry-technology-maritime-security-global-news/5453-video-japan-s-xasm-3-supersonic-anti-ship-missile-test-launch.html

Navy Recognition, "Taiwan's NCSIST Successfully Tested a Ship-Based Variant of Tien Kung III BMD Interceptor," web page, January 2, 2017. As of March 20, 2017:
http://navyrecognition.com/index.php/news/defence-news/2017/january-2017-navy-naval-forces-defense-industry-technology-maritime-security-global-news/4750-taiwan-s-ncsist-successfully-tested-a-ship-based-variant-of-tien-kung-iii-bmd-interceptor.html

Niayesh, Umid, "Iran Builds First Hypersonic Wind Tunnel to Test Missiles and Spacecraft," *Trend News Agency*, March 5, 2014. As of March 20, 2017:
http://en.trend.az/azerbaijan/politics/2249244.html

"Nitrogen Oxides; Findings from University of Calgary Broaden Understanding of Nitrogen Oxides (Nitric Oxide Chemistry Effects in Hypersonic Boundary Layers)," *Defense & Aerospace Week*, 2015, p. 56.

Norris, Guy, "Australia Pushes Toward Space with Hypersonic Effort," *Aviation Week & Space Technology*, Vol. 173, No. 15, April 25, 2011. As of March 17, 2017:
https://www.researchgate.net/publication/292519050_Australia_pushes_toward_space_with_hypersonic_effort

———, "LEA Nears Delayed Hypersonic Ground Test Milestone," *Aerospace Daily & Defense Report*, September 27, 2012a. As of March 15, 2017:
http://aviationweek.com/print/awin/lea-nears-delayed-hypersonic-ground-test-milestone

———, "Europe and Japan Forge New Hypersonic Links," *Aviation Week & Space Technology*, November 26, 2012b. As of March 20, 2017:
http://aviationweek.com/awin/europe-and-japan-forge-new-hypersonic-links

———, "Turbojet Test," *Aviation Week & Space Technology*, November 26, 2012c.

———, "Hyper Hurdles," *Aviation Week & Space Technology*, Vol. 175, No. 38, November 4, 2013.

———, "Polar Express," *Aerospace America*, web page, May 2014.

———, "Subscale Reusable Launch System Demonstrator to Fly This Year," *Aviation Week & Space Technology*, July 22, 2015.

Nuclear Threat Initiative, "South Korea," web page, April 2016a. As of March 20, 2017:
http://www.nti.org/learn/countries/south-korea/delivery-systems/

———, "Is Launch Under Attack Feasible?" web page, August 4, 2016b. As of May 2, 2017:
http://www.nti.org/analysis/articles/launch-under-attack-feasible/

ONERA, The French Aerospace Lab, "DCPS–System Design and Performance Evaluation: Projects and Research Topics," web page, undated. As of March 10, 2017:
http://www.onera.fr/en/dcps/projects

———, "Hypersonic Wind Tunnel," web page, 2009. As of March 10, 2017:
http://windtunnel.onera.fr/hypersonic-wind-tunnel

Park, Gisu, "Study of Oxygen Catalytic Recombination," 43rd AIAA Thermophysics Conference, Fluid Dynamics and Co-Located Conferences, New Orleans, La., June 25, 2012. As of March 20, 2017:
https://arc.aiaa.org/doi/abs/10.2514/6.2012-3098

Pecnik, René, Vincent E. Terrapon, Frank Ham, Gianluca Iaccarino, and Heinz Pitsch, "Reynolds-Averaged Navier-Stokes Simulations of the HyShot II Scramjet," *AIAA Journal*, Vol. 50, No. 8, August 2012, pp. 1717–1732. As of March 15, 2017:
https://arc.aiaa.org/doi/abs/10.2514/1.J051473?journalCode=aiaaj

"Photos: Hypersonic Jet Could Fly 10 Times the Speed of Sound," *Life Science,* November 3, 2015. As of March 17, 2017:
http://www.livescience.com/52681-hypersonic-skreemr-jet-photos.html

Ponirakis, Lore Anne, "Dense Core Ablative Nosetip Materials for Hypersonic Applications," *Navy Small Business Technology Transfer*, December 17, 2012. As of May 2, 2017:
http://www.navysbir.com/n13_1/N131-071.htm

"Propulsion and Power: New Findings from Nihon University in the Area of Propulsion and Power Described (Acoustic Simulation of Hot Jets Issuing from a Rectangular Hypersonic Nozzle)," *Defense & Aerospace Week*, July 2, 2014, p. 121.

"Propulsion and Power: New Findings on Propulsion and Power from Federal University Summarized," *Defense & Aerospace Week*, July 3, 2013, p. 396.

Public Law 114-328, National Defense Authorization Act for Fiscal Year 2017, Subtitle E, Missile Defense Programs, Section 1687, December 23, 2016. As of May 2, 2017:
https://www.congress.gov/bill/114th-congress/senate-bill/2943

Purohit, Jugal, "Inside the BrahMos Missile Factory," *New Delhi Mail Today in English*, February 20, 2017.

Putnam, Dean, "Ceramic-Metal Joining for Hypersonic Vehicle and Missile Components," *Navy Small Business Technology Transfer*, January 11, 2016. As of May 2, 2017:
http://www.navysbir.com/n16_1/N161-046.htm

Quick, Darren, "Scramjet-Based Project Looks to Blast Australia into Space 2015," *New Atlas*, August 10, 2015.

"Ramjet Engine for Sky Bow Sam," *Jane's Defence Weekly*, July 14, 1990.

Raytheon, "Raytheon Tech Forges the Future of the World's Largest Test and Evaluation Range," web page, August 8, 2016. As of March 17, 2017:
http://www.raytheon.com/news/feature/mobile_range.html

Republic of Singapore Air Force, "First Anti-Aircraft Guns," web page, February 23, 2017. As of March 20, 2017:
https://www.mindef.gov.sg/imindef/mindef_websites/atozlistings/air_force/about/museum/1970s.html

Richardson, Doug, "NAWCWD Upgrades Its Sea-Level Engine Test Facility," *Jane's Missiles & Rockets*, May 1, 2015.

"Russia's Lavrov Denies Meddling in European Votes, Blasts U.S. Intelligence," Radio Free Europe Radio Liberty, January 17, 2017. As of May 2, 2017:
http://www.rferl.org/a/russia-lavrov-us-trump-syria/28238716.html?utm_source=Sailthru&utm_medium=email&utm_campaign=Earl%20Bird%20Brief%2001.17.2017&utm_term=Editorial%20-%20Early%20Bird%20Brief

Russon, Mary-Ann, "Airbus Patents Hypersonic Jet That Can Travel from London to New York in Just One Hour," *International Business Times*, August 4, 2015. As of March 13, 2017:
http://www.ibtimes.co.uk/airbus-patents-hypersonic-jet-that-can-travel-london-new-york-just-one-hour-1514003

Schrijer, F. F. J., B. W. Van Oudheusden, U. Dierksheide, and F. Scarano, "Quantitative Visualization of a Hypersonic Double-Ramp Flow Using PIV and Schlieren," in 12th International Symposium on Flow Visualization, Göttingen, Germany, September 14, 2006.

Scott, Richard, "Known Unknowns: Future Operating Environment 2035," *Jane's Defence Weekly*, January 5, 2016.

Space Engineering & Technology, "Facts and Figures," web page, October 2, 2012a. As of March 16, 2017:
http://www.esa.int/Our_Activities/Space_Engineering_Technology/Facts_and_figures5;

———, "FAST20XX (Future High-Altitude High-Speed Transport 20XX)," web page, October 2, 2012b. As of March 16, 2017:
http://www.esa.int/Our_Activities/Space_Engineering_Technology/FAST20XX_Future_High-Altitude_High-Speed_Transport_20XX

"Spacecraft and Rockets; Data on Spacecraft and Rockets Reported by Researchers at German Aerospace Center (Transpiration-Cooled Hypersonic Flight Experiment: Setup, Flight Measurement, and Reconstruction)," *Defense & Aerospace Week*, 2015, p. 91.

"Spacecraft and Rockets; Researchers from University of Naples Federico II Describe Findings in Spacecraft and Rockets (Plasma Effect on Radiofrequency Communications for Lifting Reentry Vehicles)," *Defense & Aerospace Week*, April 29, 2015, p. 146.

Speier, Richard, "An NPT for Missiles?" in Henry Sokolski, eds., *Fighting Proliferation: New Concerns for the 1990s*, Maxwell Air Force Base, Ala.: Air University Press, 1996. As of May 2, 2017:
http://www.fas.org/irp/threat/fp/b19ch3.htm

Speier, Richard H., Brian G. Chow, and S. Rae Starr, *Nonproliferation Sanctions*, Santa Monica, Calif.: RAND Corporation, MR-1285-OSD, 2001. As of March 31, 2017:
https://www.rand.org/pubs/monograph_reports/MR1285.html

Steelant, J., "Hypersonic Technology Developments with EU Co-Funded Projects," *Defense Technical Information Center*, September 2010. As of March 10, 2017:
https://lirias.kuleuven.be/handle/123456789/336461

———, "Sustained Hypersonic Flight in Europe: First Achievements within LAPCAT II," 17th American Institute of Aeronautics and Astronautics International Space Planes and Hypersonic Systems and Technologies Conference, San Francisco, Calif.: American Institute of Aeronautics and Astronautics, Vol. 2243, 2011. As of March 15, 2017:
http://www.enu.kz/repository/2011/AIAA-2011-2243.pdf

Steelant, J., M. Dalenbring, M. Kuhn, M. Bouchez, and J. von Wolfersdorf, "Aero-Thermodynamic Loads on Lightweight Advanced Structures II (ATLAS II: Final Report)," European Space Agency–European Space Research and Technology Centre, October 2, 2012. As of March 10, 2017:
http://m.esa.int/Our_Activities/Space_Engineering_Technology/ATLLAS_II_-_Project_summary

"Strategic Weapon Systems," *Jane's Sentinel Security Assessment–Eastern Mediterranean*, February 7, 2017.

Svitak, Amy, and Robert Wall, "French Legislators Push Broad Missile Defense Agenda," *Aviation Week & Space Technology*, July 18, 2011, p. 26. As of March 13, 2017:
http://aviationweek.com/awin/french-legislators-push-broad-missile-defense-agenda

The Swedish Space Corporation, "Esrange Space Center," web page, undated. As of March 16, 2017:
http://www.sscspace.com/products-services/rocket-balloon-services/launch-services-esc/launch-services-esrange-space-center

Taguchi, Hideyuki, Akira Murakami, Tetsuya Sato, Takeshi Tsuchiya, "Conceptual Study on Hypersonic Turbojet Experimental Vehicle (HYTEX)," *Transactions of Space Technology Japan,* Vol. 7, No. 26, 2009, pp. 27–32. As of May 2, 2017:
http://adsabs.harvard.edu/abs/2009TrSpT...7.Pa27T

Tate, Karl, "How Europe's IXV Space Plane Works (Infographic)," *Space.com*, February, 9, 2015. As of May 10, 2017:
http://www.space.com/28456-how-european-ixv-space-plane-works-infographic.html

Taverna, Michael A., and Douglas Barrie, "Son of Japhar," *Aviation Week & Space Technology*, Vol. 169, No. 14, October 13, 2008.

Thisdell, Dan, "EADS Puts Faith in Detonation Engine," *Flight International*, 2012.

———, "SABRE Cutting Path to Live Testing," *Flight International*, 2015. As of March 16, 2017:
http://ebook-dl.com/magazine/flight-international-november-10-20159637.pdf

"TNO, RWM Create New Class of Projectile," *Jane's Missiles and Rocket*, February 1, 2007.

Tomkins, Richard, "Swedish AF Gripens Now Carry Meteor Missiles," *United Press International News*, July 11, 2016. As of March 16, 2017:
http://www.upi.com/Defense-News/2016/07/11/Swedish-AF-Gripens-now-carry-Meteor-missiles/5341468254872/?st_rec=9021468503609

Tran, Pierre, "Onera Explores Mach-8 Missile Engine Technology," *Defense News*, October 19, 2015, p. 17.

Trimble, Stephen, "LAAD11: Brazil Reveals Details of 14-X Hypersonic Vehicle," *Flight Global News*, April 13, 2011. As of March 20, 2017:
https://www.flightglobal.com/news/articles/laad11-brazil-reveals-details-of-14-x-hypersonic-ve-355517/

———, "DUBAI: China Details Performance of 'Carrier Killer' Missile for JF-17," *Flight Daily News*, November 19, 2013. As of March 20, 2017: https://www.flightglobal.com/news/articles/dubai-china-details-performance-of-carrier-killer-missile-for-jf-17-393301/

Uebler, Ulla, "Analysis and Localisation of Communications Emitters in Strategic and Tactical Scenarios," *Naval Forces*, Vol. 33, No. 5, October 2012, p. 128. As of March 20, 2017: http://connection.ebscohost.com/c/product-reviews/89635670/analysis-localisation-communications-emitters-strategic-tactical-scenarios

United Nations, Article IV of the Treaty on the Nonproliferation of Nuclear Weapons (NPT), New York, May 2005. As of August 21, 2017: http://www.un.org/en/conf/npt/2005/npttreaty.html

University of Oxford, "New National Wind Tunnel Facility," web page, undated. As of March 17, 2017: http://www.eng.ox.ac.uk/thermofluids/news-events/new-national-wind-tunnel-facility

The University of Queensland Centre for Hypersonics, "About HyShot Program," web page, undated-a. As of March 17, 2017: http://hypersonics.mechmining.uq.edu.au/hyshot-about

———, "Current Research Projects" web page, undated-b.

University of Tokyo, "Hypersonic and High Enthalpy Wind Tunnel Kashiwa Campus, The University of Tokyo," Graduate School of Frontier Sciences (GSFS) Division of Transdisciplinary Sciences, June 2006. As of March 20, 2017: http://daedalus.k.u-tokyo.ac.jp/wt/info/WTpamphE.pdf

UQ News, "Launching Australia into Space," The University of Queensland Centre for Hypersonics, August 10, 2015. As of March 17, 2017: https://www.uq.edu.au/news/article/2015/08/launching-australia-space

Von Karman Institute for Fluid Dynamics, "High Speed Wind Tunnels," web page, undated. As of March 10, 2017: https://www.vki.ac.be/index.php/research-consulting-mainmenu-107/facilities-other-menu-148/high-speed-wt-other-menu-158

———, "Lecture Series STO 234 on Hypersonic Flight Testing," web page, March 28, 2014. As of March 10, 2017: https://www.vki.ac.be/index.php/component/content/article/249-news/latest/464-lecture-series-sto-234-on-hypersonic-flight-testing

Wall, Mike, " Air Force's X-51A Hypersonic Scramjet Makes Record-Breaking Final Flight," *Space.com*, May 3, 2013.

Wall, Robert, and Michael Taverna, "Slowing Speed," *Aviation Week and Space Technology*, Vol. 159, No. 20, 2003.

Wang, Brian, "Japan and Europe Cooperating to Develop the Technology for Hypersonic Commercial Passenger Planes," Next Big Future, March 9, 2015. As of May 3, 2017:
http://www.nextbigfuture.com/2015/03/japan-and-europe-cooperating-to-develop.html

Wen, Chih-Yung, Yen-Sheng Chen, and Jiiun-Yue Chen, "Numerical Simulation of Complex Internal Viscous Flow in the HyShot-II Scramjet Model," 17th AIAA International Space Planes and Hypersonic Systems and Technologies Conference, San Francisco, Calif., April 2011. As of March 20, 2017:
http://arc.aiaa.org/doi/abs/10.2514/6.2011-2382

Wind Tunnel Technology Center, "Hypersonic Wind Tunnels," web page, undated. As of March 20, 2017:
http://www.aero.jaxa.jp/publication/pamphlets/pdf/0.5_1.27_gokucyoonsoku.pdf

Wong, Kelvin, "India Opens New Hypersonic Wind Tunnel Facility," *Jane's International Defense Review*, Vol. 47, No. 5, May 1, 2014.

"YJ-12 (CM-302)," *Jane's Air-Launched Weapons*, October 5, 2016.

www.ingramcontent.com/pod-product-compliance
Lightning Source LLC
Chambersburg PA
CBHW071513290625
28631CB00012B/17